NEW UNIVERSE THEORY WITH THE <u>LAWS OF PHYSICS</u>

by
Bobby McGehee

New Universe Theory Developed by Career Engineering-Physicist Bobby L McGehee

Cosmology and Astronomy

authorHOUSE™

1663 LIBERTY DRIVE, SUITE 200
BLOOMINGTON, INDIANA 47403
(800) 839-8640
WWW.AUTHORHOUSE.COM

© 2005 Bobby McGehee
All Rights Reserved.

No part of this book may be reproduced, stored in a retrieval system, or transmitted by any means without the written permission of the author.

First published by AuthorHouse 06/17/05

ISBN: 1-4184-9431-3 (e)
ISBN: 1-4184-9429-1 (sc)
ISBN: 1-4184-9430-5 (dj)

Library of Congress Control Number: 2005904178

Printed in the United States of America
Bloomington, Indiana

This book is printed on acid-free paper.

The <u>New Universe Theory</u> Conception !

I learned of the Big Bang theory in the 1950s. The Big Bang idea was especially intriguing and puzzling, since there are so many things about it that are inconsistent with the Laws of Physics. I have always been taught that all knowledge is based on proven and repeatable phenomena and facts that have been proven to be true. Repeatable and provable phenomena, that cannot be disproved are the **Laws of Physics**.

I am from a small town in Oklahoma, and always asked "why and how." After High School, and two years in the service, I attended college for five years, earning three degrees. (Physics, Engineering Physics, and Education). In the process many subjects were studied including Astronomy, Electron and Nuclear Physics, Electricity and Magnetism, Mechanics-Statics and Dynamics, Kinetic Theory of Gases, etc. After an exciting 32 year career as an Engineering Physicist, I returned to college for three years, studying Geology and Astronomy.

For the last 30 years I have read many scientific publications, attended Astronomy organization lectures and conferences, striving to prove (to myself) that the Big Bang is valid. But I do not find it scientifically credible. So, with more study, pondering and analysis, I arrived at a scientifically feasible origin of the universe, which from the time before it's beginning, is consistent with the Laws of Physics. I named it the **New Universe Theory (NUT)**.

Steven Hawking made the following quote; *"It would be arrogant, if not absurd, to pretend that I, single-handed could ever attempt to put together a complete picture of the nature of the universe-let alone a compelling explanation easily understood by everyone."*

What I am going to do is describe how I believe the universe came into existence, and how I believe it is continuing to develop. Reading this document requests you to consider some new, and controversial points of view. Ideas that have been taught and accepted in the Astronomy community for several decades, may be difficult to release; both old and new ideas may, and should be questioned. Please be open minded, curious, and read the whole story. Some of the terms used in this writing may be unfamiliar to many, especially non-scientific readers. So an attempt is made to make it understandable to everyone. Objects and processes described in the text, are supported by the Glossary and References. For example, all mass is known to be made of atomic and sub-atomic particles such as electrons, protons, neutrons, positrons, positroniums, quarks, isotopes, etc. Most of these particles have been known for over a century, and the rest have been known for decades. These particles will be described and discussed for those who have forgotten these since high school physics and those that have not yet been introduced. After reading this explanation of the New Universe Theory, these basic mass particles will be familiar to all readers, as will phenomena and physical relationships like "Doppler Red-Shift", "Einstein's Relativity", "Deflagration Wave", and "Velocity Enhanced Gravity", as well as some other interesting physical phenomena.

Reviewing and pondering past observations in the light of the New Universe Theory, is revealing and enlightening.

Quotes

Famous thinkers expect and encourage additional advancement of knowledge.

Astronomer ***Carl Sagan*** (1934-1996), author and presenter of the TV series "Cosmos": ***"Our descendants will marvel at the things we did not know, that are so plain to them."***
(Even though some of us are older than Carl would be today, readers of this book are those descendants).

Particle Physicist ***Steven Weinberg***, while theorizing "The First Three Minutes" of the Big Bang: ***"...even as I write this I cannot deny a feeling of unreality in writing about 'the first three minutes' as if we really know what we are talking about."*** ... ***"Will new discoveries overthrow it and replace the present (BB) standard model with some other cosmology?"***
(I respectfully present this book for consideration).

Author of astronomy books (including the coffee table book "Galaxies") ***Timothy Ferris*** : ***"The cosmological theories of today may be looked upon by our descendants with respect, bemusement, scorn, or even hilarity ..."***.
(With respect, Timothy, I agree that all four are appropriate).

"I have great respect for those who invested so much innovative thought seeking the explanation for the origin of the universe. Now with eager pride, I present the 'New Universe Theory'."   ***Bobby McGehee***

Purpose

To reveal the New Universe Theory (NUT) concept and substantiate how the Universe is generated from Primordial Matter by processes compliant with the Laws of Physics. My objective is to stimulate discussions about the origin of the universe, and thereby stimulate more interest in the exciting field of Astronomy and Cosmology, and to promote the use of scientific principles.

A fresh start with the Laws of Physics:
1. **Exposes the Big Bang (BB) idea as an inaccurate hypothesis.**
2. **Reveals the "red-shifts" misinterpretation which led to wrong conclusions.**
3. **Provides a "New Universe Theory"; Primordial, Processes, and Matter.**
4. **Suggests the size of our Universe is larger than ever before imagined.**
5. **Show Our Universe is not exploding to death, but is growing with vigor.**
6. **Promotes Astronomy interest and understanding as a science of physics.**

Readers will be first to know how the world will be astounded !
As a Career Engineering-Physicist, I have pondered and questioned for several years, the 'accepted' cosmological concepts, as did revered British astronomer Sir Fred Hoyle, and pioneer radio astronomer Grote Reber. After scrutinizing accepted theories and their inconsistencies with the Laws of Physics, I arrived at this new provable concept and revelation. The New Universe Theory is so interesting and exciting! I have provided background information and related knowledge for all to comprehend, understand, and get the NUT into perspective. This book is structured with the hope that the intriguing NUT will be understandable for all curious minds.

Dedication

To my wife, our children, and my parents.

Thanks to Nancy Ruth (nee Williams) from Hugo Oklahoma, my most ardent supporter and life partner since 1951. This book is dedicated to Nancy and our two children. Daughter Sue, with son-in-law Jim Hart; and our son Bobby Jr. Also to Granddaughter Elaine Hart whose future is just now arriving.

Thanks for the philosophical guidance from both my Mother, Mary (nee Hollis) and my Father, William E. McGehee. They are more appreciated every day even though they passed away years ago, in 1973 and 1975.

Family and friends continue to be a vital source of encouragement and support.

Contents.

The New Universe Theory (NUT) describes Primordial Matter and the Universe generating processes in compliance with the Laws of Physics.

Front Material v
                 The New Universe Theory Conception! v
                 Purpose ix
                 Dedication xi
                 Contents. xiii
                 Abstract xv
                 Foreword xvii
                 Synopsis: xxi
                 Preface xxiii
                 Premise xxv

Chapter 1 ... Introduction 1
Chapter 2 ... Previous Theories 15
Chapter 3 ... Red-Shift Mis-Interpretation 25
Chapter 4 ... Big Bang and the Laws
                 of Physics 41
Chapter 5 ... Analogies of the N U T 47
Chapter 6 ... New Universe Theory 59
Chapter 6-I ... Stage I The Proto Universe 63
Chapter 6-II ... Stage II Reduction Mechanisms 85
                 Stage IIA 86
                 Primordial Matter Transforms to Energy
                 Stage IIB 92
                 Energy Transforms to Mass
                 Stage IIC 98

	Particles Coalesce Into Observable Objects	
	Synopsis of Sub-Stages II A, II B, plus IIC	112
Chapter 6-III ...	Stage IIICentral Universe	117
Chapter 7 ...	Entropy & Deceleration	141
Chapter 8 ...	Fluid Flow	167
Chapter 9 ...	Size & Age of the Universe 20th Century Version	185
Chapter 10 ...	Size & Age of the Universe 21st Century Version	195
Chapter 11 ...	Questions & Answers& More Questions	217
Chapter 12 ...	Conclusions	233
Epilog:		243
Glossary Terms and Definitions		245
Appendix 1.0	"Laws of Physics"	267
Appendix 2.0	Numerical Values	273
Appendix 3.0	Equations & Descriptions	277
Appendix 4.0	Bibliography	285
Appendix 5.0	Acknowledgments:	293
Appendix 6.0	Author's Credentials'	295
Appendix 7.0	Tributes:	299
	Hermann Bondi 1919—20__ __.	
	Thomas Gold 1920—2004.	
	Fred Hoyle 1915 —2001.	
	Grote Reber 1911 -- 2002.	301
Appendix 8.0	Concluding Comments	303

Note To Readers:

Over the years, I have found that by previewing the text in advance of starting classes in a new study course, enjoyment and understanding of the new subject was greatly enhanced.

I recommend that this book first be perused by viewing all Figures (Graphs, Tables, and Photographs) in the sequence presented, reading the figure captions as you progress. Then proceed to read through the entire book. When you find some of the material you do not fully comprehend, you should just keep on reading to get the rest of the story. It is more important that you get the overview of the New Universe Theory. You can then return to study the material you did not initially comprehend as well as to further study phenomena explanations which you wish to understand in more depth.

This book has been written with the intent of minimizing the 'fog factor' by as we go explaining non-common words and terms. Most of these terms are already known to physicists, and these terms will become familiar to all as we read on.

Foreword

After giving a briefing on the New Universe Theory, the following unsolicited note was sent to me:

As humans, we are limited to the workings of the human brain. Our brains perceive the world through assumptions based on previous experiences and knowledge. Brilliant people throughout history have developed an assumed understanding of the universe based on the findings and theories of the great minds that have preceded them. It is comforting when explanations are based upon the solid ground of long accepted assumptions of truth.

When all of the pieces of a puzzle fit neatly together, we feel a sense of peace and security. Perhaps this is why so many have struggled to fit the questions surrounding our observations of the universe into the already formed puzzle frame called the Big Bang. It is uncomfortable and frustrating to undo what has already been done, and start again. But, sometimes when we do start fresh, and pick up the pieces of the puzzle already placed, we find that they didn't fit at all. In fact they were forced into place by shaving the corners. Now we can put it together correctly and the pieces we were having so much trouble with fall neatly into place.

Bobby McGehee has done just that. He has been looking at the Big Bang puzzle since it was put on the table and has never been satisfied with how the pieces fit together. So he instead looked at the individual pieces of the puzzle, redefined the frame, and found that they do fit together without adjusting any of the Laws of Physics to make them fit into an assumed shape. Now the puzzle of Generation and ongoing growth of the cosmos is tidy and the pieces fall neatly into place. As people come to understand this New Universe Theory, they will feel comfortable and secure once again because all of the pieces of the puzzle fall neatly into place without trimming any of the corners.

Foreword by Suzanne (Sue) Hart

Nancy McGehee, Author's Wife, made the quote:

*"When a long term belief is taken away, it must be replaced
with something of equal or greater value."*

This book presents both; by Revelations and New Universe Theory.

Synopsis:

My **New Universe Theory (NUT)** is totally compliant with the **Laws of Physics;** It is a theory as to how the universe could have been and is continuing to be generated. Because the NUT is compliant, the NUT answers questions raised by the inconsistencies between the Big Bang (BB) and the **Laws of Physics.** The **NUT** identifies why the BB was not a credible concept, and the BB could not happen with the laws of physics. I hope and believe that after reading this book, the reader will agree with me that the BB did not happen and the **NUT** certainly did!

Until now, no other acceptable explanation has been conceived; it was thought any new idea had to comply with the 1920's red-shift mis-interpretation. My fresh new look at the red-shifts vs distance reveals a scientifically logical interpretation, which leads us to the technically feasible**, and Laws of Physics compliant, New Universe Theory.**

Inner-sanctums of our universe's Manufacturing and Production operations are revealed in this presentation of The New Universe Theory. This document describes the fascinating and enlightening processes with readily comprehendible analogies; providing understanding for all levels of technical backgrounds. Described is what I believe is correct red-shift interpretation followed by descriptions of "Matter Reduction Mechanisms" from the "Deflagration wave".

This New Universe Theory reveals how the universe is not expanding to its death, but shows that the universe is continuing to grow with vigor. Also described is a technically valid concept for Primordial Matter, which includes space, mass, and energy in the form of possibly an infinite size monolithic-hexahedron-crystal, made of uniformly distributed positroniums, at the same average density as the observable universe. Yet, proto-matter in proto-space, and its counterpart, the universe, on an average, are more sparsely occupied than the best vacuum achievable anywhere on earth.

The NUT phenomena is a continuing process, reducing "primordial matter" and <u>adding</u> the processed matter as mass, energy, and space ***onto*** the outer surface of our universe. Re-examination of previous and recent observational data in light of the New Universe Theory reveals new valid ways to estimate the size and age of the universe. It is much older and larger than previously believed. The universe is continuing to grow by the transformation of primordial matter. We know it is not being created, because by the Laws of Physics, "matter can neither be created nor destroyed."

It is anticipated and reasonable to expect that initially, the astronomy community may not accept the NUT with open arms. However, after this New Universe Theory is questioned, studied, and understood, I expect it to be enthusiastically and 'universally' accepted throughout the Physics, Cosmological fields, and then by other scientific and astronomy communities. The NUT should be questioned and critiqued by both amateurs and experts, and then modified and refined to accommodate new findings and substantiated valid constructive criticisms. It can then be further developed in detail as well as expanded to be more comprehensive.

Preface

This work has been developed and written without benefit of colleague consultation, peer review or referee prior to manuscript draft. This work is not intended to step on toes, ruffle feathers, or embarrass those who have made claims through some of their past statements in support of, or through deductions from the faulty BB concepts.

As recently as the December 19, 2003 issue of the prestigious publication journal, "Science" by the American Association for Advancement of Science (AAAS), the editors named the 'proof' of the existence of 'dark energy' as the 'Break Through of the Year'. Now, the New Universe Theory (NUT) reveals that 'dark energy' is an erroneous hypothesis. BB supporting deductions often provide what appears as valid data, but is being interpreted erroneously. The editor of Astronomy magazine states on the cover of the December 2002 issue "there are 5 reasons why you should believe the Big Bang". Actually, all five reasons better support believing the NUT. In the same issue, the feature article, by Switzer of DePaul University, he states that a recent National Science Foundation survey found that <u>2/3 of the people in the U.S. do not accept the BB.</u> This New Universe Theory validates the 2/3rd's opinion. The NUT is consistent with the definition of 'science', and it has been developed totally compliant with the Laws of Physics, and acknowledges <u>the Laws have always been valid</u>. This New Universe Theory concept is explained comprehensively in the ensuing chapters.

I have made some extrapolations from the New Universe Theory to arrive at some deductions, which are technically valid, but the conclusions are known to not be precise. Some of the data needed for input to the New Universe Theory 'model' are not yet available; tentative estimates and assumptions are therefore made. Now, after this debut, the New Universe Theory 'model' needs to be fed proven and corrected data from astronomy observations to arrive at more refined values and authenticated conclusions. With the NUT, I have, in the text, addressed a few of mankind's persistent questions, and more are addressed in the Q&A Chapter: Where is the center of the universe? What is the visual diameter of the universe? What is the real diameter of the universe? Where are we? Are there primordial matter alternatives? Universe's age? Is there an end? What is the source of so much angular momentum? And more.

Premise

Progress in science is accomplished when something new has been observed and then is recognized and interpreted within the framework of the **Laws of Physics** (Appendix I). Progress then continues through recording of relevant findings followed by in depth thinking, analysis, and deductive reasoning, while continually reconciling all ideas with proven facts and the Laws of Physics. Tentative explanations are called hypotheses, which if not disproved by comparison with the Laws of Physics, are further substantiated by experiments and observations, can then be described as theories. If theories are proven inconsistent with the Laws of Physics or are otherwise incorrect, they must be cast aside in favor of new ideas, or modified to accommodate conflicting observations. Analysis must always involve careful examination of the data and interpretations in the light of established scientific laws.

Physical Laws are general statements that summarize phenomena and processes that have been substantiated by many experiments and observations. Basic laws cannot be simply declared, temporarily suspended or ignored, either initially or after cursory analyses. Conclusions from observations that are in conflict with scientific laws must either be recognized as not valid, or the law must be modified to accommodate any new valid conclusion.

Extending knowledge and furthering understanding of Cosmology is the intent in this document; through my re-interpretation of observed red-shift and through the revealing of my New Universe Theory for the origin and generation of the universe. New hypotheses and theories are developed with new interpretations of past observations, with the use of logic, and by using established scientific facts and data. New hypotheses have been considered for inclusion only where the interpretations and analyses are consistent and physically possible within the **Laws of Physics**. By presenting the New Universe Theory through this document it is available without censor or filtering for review by all, and can be critiqued by professionals, students, amateurs, and any who are readers of scientific information. Critiques and questions by professionals, semi-scientific, and lay readers are not only welcome, but are sincerely solicited. All views will be respected and carefully considered.

It is recognized the basis of communications require consistent interpretation of terms. It is the intent to make this writing readily comprehendible for readers from all disciplines and with all levels of scientific background. Terms, some common, that too often have been misused or have multiple definitions, are included in the glossary, along with some not so common scientific terms. In the Glossary are terse definitions of terms as they are intended for use in this document. To set the scene, a few are stated as follows: **Science** is defined as the study of dealing with facts and truths. Facts being provable by valid verifiable observations, and truths being primarily the **Laws of Physics**. "Belief without proof" is contrary to scientific thought as 'belief' can be accepting something as valid in the face of no verifiable observations, and therefore no proof, even though there may sometimes appear to be some valid evidence. **Knowledge** is belief only with proof. For example, Astrology is fun, but it is typically one of many pseudo sciences which are not valid as science, as they are simply man made, relying on ideas with no proof or even tangible evidence. Likewise, all myths are also man made and cannot be proven to be true, so therefore are not valid by simple application of proven knowledge with logic and deductive reasoning. **Matter** is not a mysterious substance; Its constituents are both mass and energy. The basic **Laws of Physics** have been repeatedly proven and never disproved, therefore must continue to be recognized and respected as long as they are not shown to be invalid. There should always be a lingering doubt about any claim or interpretation from experimental and observed data; scientists should always be careful to not confuse 'obvious' interpretations, with observed and proven fact. Progress in science has too often been set back and slowed because of acceptance of erroneous theory and hypotheses as fact.

Yet, having dwelled on the above, the mainstream astronomy community has for several decades have become victims, succumbing to the comfortable trap of being in the popular compliance majority. Unfortunately, the Big Bang hypothesis has been allowed to exist far too long. And now a revelation; the New Universe Theory, the NUT! It is my opinion that technical papers on astronomy or any other subject should not be accepted for publication by technical societies unless all applicable Laws of Physics are addressed.

A Special consideration in writing this work:

I have been advised by writing experts and publishers that for every equation encountered in the text of a book, one half of the remaining readers are lost. In an effort to have the complete concept of this **New Universe Theory** available and comprehendible to all, detailed equations are relegated to the appendix and are avoided in the text. Analogies are frequently used as are nomographs to allow technical comprehension without extensive analyses. When the physical deductions require mathematical analysis, the relationships are written in words, and the deductions are presented in graphical or table form so that the concepts can be comprehended by pondering the related graph or table. This is my chance to get back at some of the authors of my college day technical text writers when they would present a technical statement and then add; "Proof is left for the student". In this work, I still consider myself a student and will be presenting the technical concepts with enough reference to the Laws of Physics that, if the reader wishes, analyses can be accomplished to substantiate claims. The verification of conclusions will this time, be left to the professional, text book writers, as well as students. (So there). Applicable math equations are relegated to the appendices.

Chapter 1 ... Introduction

There are literally thousands of astronomers and cosmologists, both professional and amateur, who can describe the observed and observable universe in a manner that is elegant, accurate, and clear. My purpose in this document is to disclose my New Universe Theory (NUT) that explains the cosmological processes, which includes how I believe all of the universe and its contents came into existence. Also described is what I believe is the source of the matter from which the objects are made. This NUT exposes many new revelations, and it challenges the total 20th century Big Bang (BB) concept. The aim of this writing is to be totally scientific and credible, staying feasible and plausible within the boundaries of the Laws of Physics. It is now time for this scientifically creditable cosmological revelation.

Pathfinder Scientists who promoted integrity of knowledge command respect and admiration because they never compromised their scientific standards, even in the face of ridicule and cruel punishment: Hapatia of Alexandria, Galileo Galilei, Fred Hoyle, and Grote Reber, among others. Hapatia of Alexandria is credited with saying "wrong knowledge is worse than not knowing". Religious bigots executed Hapatia by the scraping of her flesh from her bones with cockle shells; To hide her superior knowledge, her library was burned and that started the Dark Ages (some of the dark ages policies continue today!); Galileo was sentenced to life imprisonment for proving that the center of the universe was not Roma, or Earth; Sir Fred Hoyle was openly ridiculed by other astronomers because he would not accept the Big Bang as the origin of the universe; Grote Reber, pioneer of radio astronomy, also is not given due respect because he openly said the Big Bang is Bunk. (BBB)

Revelation of this scientific New Universe Theory and related phenomena while emphasizing the Laws of Physics, is intended and expected to arouse skeptical interest and to be a learning stimulant for those, both young and old, with curiosity about scientific phenomena, facts, and their interdependence. As long as we continue our 'student' desire-to-learn mentality, life continues to be filled with enlightening dynamic revelations, vitality and excitement. Knowing more about cosmology may not change the life style or comfort of most of the life forms on earth, but intelligent beings are distinguished by a natural curiosity, desiring to learn and explore. Surely, mankind has evolved for more purpose than to just utter bewildering words such as 'who cares'. Hopefully, this writing will stimulate more curiosity and eagerness to learn. New and exciting scientific phenomena are all around us, and when something really great is revealed or achieved by those who pursue cosmological knowledge, something even larger than the universe is provided..."All of us with awareness <u>are</u> the Consciousness of the universe".

The size of the universe is almost beyond belief. It is huge, beautiful, and fascinating. Figure 1.1 artist's pictorial description is of only the minute segment of the universe in which we live. This view is a billion light year (radius) which encompasses a volume of space centered on earth;

the artist appropriately describes this sphere as "The <u>Local</u> Universe".

The concept of the New Universe Theory reveals that a similar volume of mass and space is being added to the periphery of the universe every hundred years or so. It is not just an expansion of the existing space that contains only past generated matter. The New Universe Theory, which explains how the universe started 36.9 Billion years ago, and how it continues to grow.

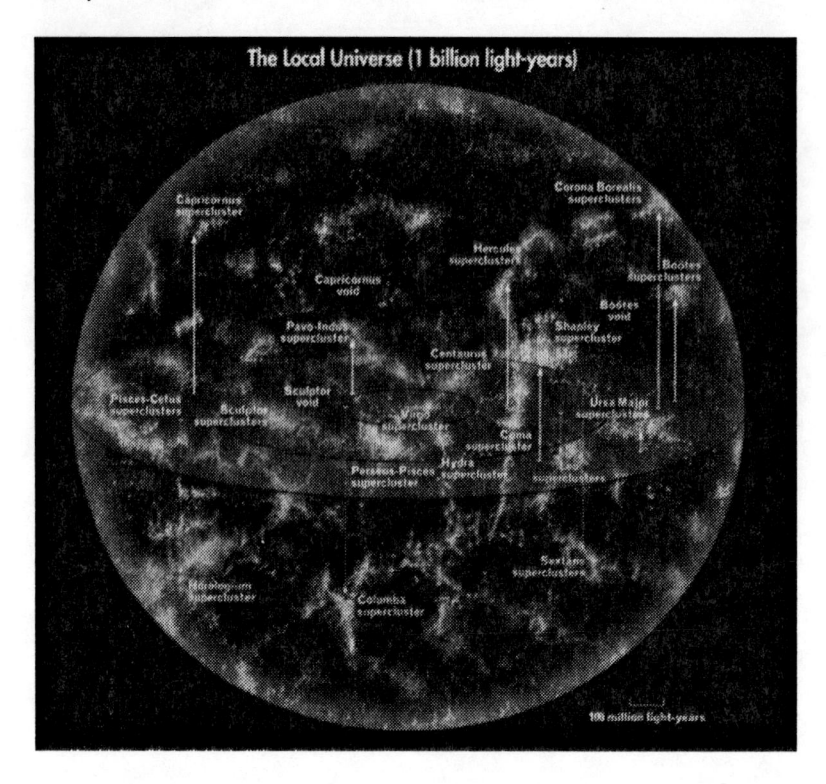

Figure 1.1. Local Universe. Earth and our sun's planetary system are embedded within the Milky Way galaxy which is deep within the Virgo Super-Cluster at the center of this map. This work (by Elisabeth Rowan of Astronomy Magazine) was constructed as a one billion light-year (BLY) radius sphere. A larger galaxy grouping, Abell super-clusters, are not shown. They extend through the periphery of the local universe, and are scattered from .5 to 3+ BLY, in approximately in the opposite direction from the Bootes super-clusters. Map was originally published with an article by free-lance writer Steve Nadis. (Printed here with permission.)

Astronomy is defined as the science, primarily observational by both direct and indirect means, that includes study of the material universe beyond planet earth. *Astrophysics* is the science that is involved with the physical properties of celestial processes and the interaction between matter (*mass and energy*) both in the interior of celestial bodies and in outer space. *Cosmology, a science,* is the physics that deals with the origin and general structure of the universe, including its smallest particles, elements, components, and laws, and especially with such characteristics as matter, space, and time. Cosmology is also involved with the evolution of the universe. *Physics* is the science that deals with matter (*by definition, matter includes both energy and mass*), relativistic effects, motion, and force. *Science* is the branch of study dealing with logic, natural laws, facts, and truths, systematically arranged and showing the relationships and operations of physical phenomena. *Logic* is the branch of the science of philosophy that deduces conclusions governing correct and reliable inference ..... it is a particular method of reasoning or argumentation. *Math* is a method for expressing logic. *Fact* is something that actually exists; it is reality ....something known to exist or to have happened. *Truth* is the actual state of conformity with fact and reality. *Laws* of science are descriptions of phenomena and processes that have been proven, are repeatable through experimentation and observation, and cannot be disproved. Having considered, verified and clarified the preceding terse definitions with several dictionary and encyclopedia references, it is appropriate to proceed with the explanation of the deductions and reasoning that concludes with the **'Millennium 3 New Universe Theory'** for the origin of the universe. Analogies are frequently used to readily convey new concepts and thoughts to readers at all technical levels. So that all who read this work will understand the intended meaning of all used terms and phrases, it is recommended that, early on, the glossary be perused to assure thinking continuity to written meanings.

Thinking Back

After reading about the Big Bang idea, back in the early 1950's, that concept's inconsistencies with the basic laws of physics immediately raised doubt and questions in my mind, as it was with many others. It has continued to be puzzling and nag at scientific logic. In the era of the 1920's through the 1950's, several "origination of the universe" ideas were proposed but most were dismissed as inconsistent with science. The steady state continuous creation was postulated by Hermann Bondi, Thomas Gold, and Fred Hoyle, and that idea was publically supported by famous British astronomer Sir Fred Hoyle, but it was later proven to be impossible in view of the inadequate amount of energy available in interstellar space to support it. *(They were on the right track, but they quit thinking too soon).* The Plasma Theory was proposed by Swedish astrophysicist Hannes Olaf Gost Alfven, but it failed for lack of evidence to support it. The only apparent choice for consideration was limited in most minds, to the big-bang because of the "too-obvious" (misunderstood and mis-interpreted) evidence, which many called proof!

Now, it has been many decades since the Big-Bang hypothesis was invented and it continues to be allowed, even embraced by many astronomers and writers for the astronomy community. Yet it does not fit within the framework of known and proven Laws of Physics. In recent years, the big-bang has continued to be crutched with other supporting hypotheses contrived as "inflation", "multiple bangs", "multiple dimensions", "tired light", even light waves deteriorating with "entropy", and most recently (in 2003) by a theory submitted by Joao Magueijo in a book about "Faster than the Speed of Light". All of these explanations are well intended, but are in conflict with the laws of physics. Some have preposterously said the Laws of Physics of today didn't apply back then, or were different in the early universe. Numerous writings have been published which include lengthy verbal, pictorial, and mathematical explanations. Extensive computer modeling has been developed in support of the faulty BB. Computer models use math which is a way to express logic, but even though logic may use valid processes to prove a conclusion, the conclusions are not valid if the initial assumptions are incorrect.

It was in 1948, after the fall semester at Oklahoma State University when time allowed me to spend a couple of days with my parents in Enid before going back for the spring semester. When describing to my mother that one of the preceding semester's interesting courses had been Philosophy 101, which was actually a course in simple logic. My mom was concerned and spoke up saying, "Oh Bobby, I wish you hadn't studied logic because that will cause you to lose your religion". I explained to her that it had nothing to do with religion, it dealt with deductive reasoning from specific facts. That conversation continued, but on other topics. She was a 'bible belt' life long religious person and had only an 8th grade education, but, thinking back, many of her philosophical statements continue to astound me, even after so many years, as to her in-depth wisdom and understanding. Thanks to Philosophy 101 and the application of logic patterns of thinking, this may be what is allowing me to wash away the 20th century Big Bang zealous ardor of many astronomers and cosmologists. The BB does not comply with logic or the Laws of Physics. It is time for us to discard old invalid concepts even if they are what we have been taught to believe and be comfortable with in the past. The New Universe Theory presented in this work meets all of the standards for displacing and replacing the old BB.

In attempts to explain some of the inconsistencies of the BB, some in the Astronomy community have invented other concepts, like the afore mentioned 'inflation'. With respect I admire an intelligent young astronomer, Rogier Windhorst, from the Astronomy Department of Arizona State University, who in 2001 gave an informative lecture to the Sun City West Astronomy Club, on the subject of inflation, none of which I could accept. Rogier, appropriately, credited well intentioned Alan Guth for the 'inflation concept'. But most importantly, Rogier inspired me to get off my hind quarters, proceed writing to describe my red-shift interpretation and develop my related 'New Universe Theory' for generation of the universe. (Which totally eliminates the supposed need for inflation). I considered inviting Rogier to collaborate on this writing, but I assumed Rogier would not have adequate time. I decided that to tell the story in context, which required that it first be put in writing. And then, all readers could digest the new red-shift interpretation together with the New Universe Theory.

The only person whom I felt certain would be both knowledgeable and totally receptive enough to listen and hear me out would be Fred Hoyle, a well known British astronomer who reserved his support for a (then) yet to be developed scientifically feasible and plausible concept. He often stated he could never accept the Big Bang idea that requires the universe with all of its matter and space be spouted in all directions, out of a single point with no dimensions. He always had a fierce dislike for the Big Bang. I looked forward with childish fantasy to presenting him with my New Universe Theory. Unfortunately, my timing was bad. I later learned Fred Hoyle had passed away in August 2001, while I was just thinking of putting my thoughts to paper.

Nuclear particle physicist Steven Weinberg wrote a book devoted to explaining the physics of the 'model' of the Big Bang. I have been told George Gamow was actually the primary developer of the physics sequences for the BB. After a recent (re)reading of Steven Weinberg's book "First Three Minutes", I now recognize that Weinberg shows underlying scientific credibility, when, in his first book release and in his own words, **Weinberg asks *"...Will new discoveries overthrow it (big bang) and replace the standard model with some other cosmogony? Even revive the steady state model? Perhaps. I cannot deny the feeling of unreality in writing about (the BB)."*** Even with this admission, he spent much effort lecturing, writing, rewriting, and later publishing revised updates describing the way, how and why, the big bang was theorized to have occurred. A copy of his work is recommended for the personal library of every person interested in cosmology, or just the history of mankind's thinking.

This New Universe Theory opens many areas which require development of supplemental concepts, theories, and analyses. But that is how knowledge works, ...curiosity and analysis that continue result in follow-on concepts to displace, modify, or add to older ideas. As ideas further develop, the results will lead to even more understanding and new knowledge. Any and all new follow-on concepts must also totally agree with the Laws of Physics to be credible and to even be considered.

Forward Thinking

The 1920's observation that the farther away an astronomical object is from us, the greater the red-shift, is not disputed. The shift indicates that objects are moving <u>forward</u> and those farthest away are separating themselves from us at a faster rate than the ones closer to us. This is undeniably valid. As the light emitting galaxy moves away, the wave lengths of that galaxy's composite light from its light radiating stars is stretched to longer (redder) wave lengths. The faster the receding galaxy's velocity, the more the stretch of the galaxies' emitted light waves. However, <u>red-shift</u> is a measure of velocity of an individual light source, and <u>is in no way an indication</u>

<u>of acceleration</u>. **The idea that the universe is expanding with acceleration was strictly an interpretation, that I believe was erroneous. Additionally, I believe another wrong assumption is that by backward extrapolation from that wrong interpretation, everything appears to have come from a single point source and time. That theorized 'event' was derogatorily tabbed by Sir Fred Hoyle as the big bang. In Grote Reber's words, the "Big Bang is Bunk". (BBB)**

Hubble's and Humason's place in history is assured for two significant confirmations which they achieved in the 1923 and 1929: <u>They confirmed V. M. Slipher's discoveries, which possibly are the greatest in the history of science</u>. First; In December of 1912, V. M. Slipher of Lowell Observatory at Flagstaff Arizona, Slipher observed Andromeda is in fact another galaxy, proving that we live in a universe which is much larger than the Milky Way Galaxy; and <u>Second,</u> confirming Slipher's 1913 observation of increasing red-shift with distance. Carl Sagan stated, (ref 9) "Of the Humason-Hubble team, Humason was the expert spectral analyst, but Hubble boldly declared the distance relationship was Hubble Law". Their claimed observation about universal "expansion" is partially correct, but was an incorrect interpretation as it is valid only in a manner of growth, and completely <u>opposite</u> from acceleration as Hubble imagined and stated..... I believe that erroneous thinking, as accepted by most astronomers to date, has misled the astronomy and cosmology community for the remainder of the 20th century and now it continues into the 21st century, into millennium #3. With what I believe as 'erroneous accelerating expansion' interpretation of red-shift, many competent scientists have conducted credible research but then misinterpreted other astronomical observations as well. I am pleased to reveal the opposite interpretation, which is scientifically feasible within the Laws of Physics, and therefore leads to a plausible (to me) scenario for "Generation, (<u>not creation</u>) of the universe".

Objects near and far, are generally traveling in directions away from us and from each other as shown by the observed red-shift; however, they are <u>not accelerating</u>, all <u>objects are in fact decelerating</u>. You cannot simply connect the dots! The near objects are not going in a path to where the more distant objects are, have been, or are going; they are simply just moving, in the same general direction as are their nearby neighbors. The nearer to us objects having decelerated over a longer time period, are traveling at lower speeds due to increasing entropy, scientifically explained later in the NUT.

Until now, the only choice for continued consideration for origin of the universe was limited in most minds to the big-bang because of their mis-interpretation of the red-shift evidence, which was "so obvious", many called that proof. The speed of objects (galaxies) that are farthest away, moving faster than the objects not so far, was interpreted to demonstrate and "prove explosive expansion".

To assure that we don't ever again fall into the 'accept it on faith' trap, all assumed knowledge should be reexamined from time to time as related facts are discovered, and as other deduced knowledge is acquired. Re-examination and serious scrutiny of the Big-Bang concept has Resulted in my development and invention of the New Universe Theory.

Timothy Ferris in his intriguing coffee table book on "Galaxies" states; *"The cosmological theories of today may be looked upon by our descendants with respect, bemusement, scorn, or even hilarity!."* I believe Timothy is correct to expect this, not only by our descendants, but even by ourselves as well. We now can look at the big bang and supporting hypotheses with all of his descriptive terms. None of us stood up and protested early on. Then, we did not then have a replacement concept. Not until now! Yes, now we do have one, and it is exciting and provides a revolution in thinking.

I sincerely apologize to those who are now deceased, especially British Astronomer Sir Fred Hoyle, Thomas Gold, and Pioneer Radio Astronomer Grote Reber, (tributes in the appendix), for my not developing this New Universe Theory in time for them to review. I am a career engineering-physicist, and now I am also an amateur cosmologist. Myself, and most other professional and arm chair 'scientists' appreciate Hoyle and Reber, contributors to mankind's technical comprehension and understanding of the universe. Their legacy: "Be Skeptical".

Many people outside the technical fields and even a few that are directly involved do not recognize how serious and stringent the efforts are in Scientific and Technical Societies to maintain credibility for their organizations. All papers that are accepted by most societies are required to comply with strict standards and are judged by preview and review boards for scientific credibility before the paper is given consideration for publication. Written articles submitted to semi-technical magazines and other commercial publications are of course previewed and prejudged by their editors. All new ideas should be aired for consideration, but surely, all technical papers on new concepts or follow-on ideas and analyses should receive preview that assures at least the Laws of Physics are addressed, before they are given any serious consideration for acceptance for publication. Hopefully, this New Universe Theory will allow thinking people of non-scientific and scientific backgrounds alike, to understand and appreciate the real universe in which we live. Much exciting work is yet to be done, and many new discoveries lie waiting to be found.

Note to my readers: My Professor of Electron and Nuclear physics, the late Dr Frank Durbin of Oklahoma State University, told me that if we students present the correct explanation for the concept of a problem solution,

he would know that we understood the physics of the related phenomena. He said he was teaching physics, not arithmetic. He further said that anyone can plug the numbers into the equations and operate a slide rule or calculator. He would give 90% credit for the solution, the accuracy of the answer was only worth the other 10%. I always admired him and respected his philosophy. This New Universe Theory is presented with his philosophy in mind which spares the reader of calculations that are not necessary for text comprehension. For those who wish to pursue analyses and get more than a 90% grade, some of the applicable equations are presented in the Appendixes.

Readers may find some of the analytical results in the New Universe Theory that are not precisely accurate, but I believe in writing the NUT that it is more important to get the NUT "out-there" for enjoyment of my revelation as well as for the scrutiny of conclusions. I think Professor Durbin would give me at least 97% credit for the NUT concept. On the other hand, my professor for English Composition would probably give me a C-minus. I met my wife-to-be in her class, and as most understand, my attention was not on learning to compose. I will anxiously look forward to hearing from readers and reviewing their constructive comments.

Chapter 2 ... Previous Theories

Big-Bang.... It was invented and accepted by intelligent well-intentioned astronomers, from the 1920's to the present, attempting to fill a void in the understanding of our universe. It was, and currently is identified as a theory...Some have said, even written, that it is fact ... but in my opinion, the idea should never have been described as other than, a hypothetical concept. The following definition quotes are listed for illustrating the paradox in thinking.

*From: "Webster's Encyclopedic Unabridged Dictionary": Big-Bang Theory, ..."*a theory that deduces a cataclysmic birth of the universe from the observed expansion of the universe, cosmic background radiation, abundance of the elements, and agrees with the laws of physics." (Ref 1). *(The part of this statement in the encyclopedic dictionary that says "...and (agrees) with the laws of Physics" is erroneous).*

*From: Astronomy Text "Exploring the Dynamic Universe" by Theodore P Snow: Big-Bang, ..."*a term referring to any theory of cosmology in which the universe began at a single point, was very hot initially, and has been expanding from that state since." (Ref 2) *(Technical feasability was not addressed)*

Also, From the Astronomy Text by Snow: Cosmology, ...
A term that is technically, the study of the universe as it now appears, ... The study of its origins is cosmogony, and this word applies to the big bang as well as to all other theories on the origins of the solar system, once thought to be the entire universe. However, in practice today the general subject of the nature of the universe and the physics of its evolution are lumped under the heading **Cosmology**.

Study of the science of cosmology, together with the historical development of this intriguing field of Physics is possibly the most intellectually maturing endeavor mankind will ever have. Mankind continues to mature with the quest for more understanding and knowledge. Acquired new knowledge must be comprehensively audited together with past knowledge and then revised to comply with the laws of physics. Paying close attention to interpretations of observations and data. A cursory review of the origin of the Big Bang theory shows how our knowledge has evolved, and how an erroneously conceived hypothesis has been accepted as fact due to mis-interpretations have been allowed to stand, which are clearly outside already established laws of physics. Advancement of science suffered only because we quit searching for solutions that are totally compliant with the Laws of Physics.

Humason and Hubble confirmed Lowell Observatory's V. M. Slipher's decade earlier observations associated with Hubble's name. <u>First</u> was Slipher's proof that the Andromeda nebula was in fact another galaxy similar to the Milky Way. Galileo before them proved N. Copernicus's theory, showing that the solar system "world" was not centered on Rome, or even the earth as previously claimed by C. Ptolemy. Hubble and Humason confirmed Slipher's discoveries that the Milky Way Galaxy is not the center of the universe, since other galaxies exist. They also examined Slipher's data and confirmed that Red shift increases with

distance, another significant observation. Later, it was claimed that everything in the universe <u>not only originally was, but also is still</u> located at the point of origin of the BB, (everything is expanding away from everything else). Now however, the New Universe Theory shows that the BB is not valid and we are a long way from the center of the universe.

Red-shift is the shifting of the light from distant 'receding' galaxies toward the red, longer wave length end of the visible light spectrum. As the light source (galaxy) speeds away, the wave lengths of light are stretched and elongated in relation to that speed. This is known as a Doppler effect. So, shifting in the red direction indicates that the light source and the observer are moving at some velocity away from each other. The result is elongation of the observed light waves. Since light travels at the same speed everywhere (almost), the wave lengths are lengthened precisely in relation to the speed of separation, in accordance with the Doppler equations. (Included in appendix)

The discovery of an increase of the red-shift with an increase in distance was almost immediately (mis)interpreted and accepted by most astronomers, to imply that the universe is expanding with the galaxies flying apart at ever increasing (accelerating) speeds.

The exploding universe hypothesis took shape earlier in the mind of Georges Lemaitre, a Belgian cosmologist and Catholic Priest. In 1927 Lemaitre revealed his theory that the universe started out very small and expanded enormously.
Also, Hubble claimed that this continuing (<u>apparent</u>) universe expansion was at a constant rate of 'increase in velocity' (which is acceleration) compared to distance. Astronomers speculated and concluded by reverse extrapolation that Humason's and Hubble's erroneous interpretation "proved"

the universe had originated from a single point at a single instant in time.

In 1948, George Gamow and Ralph Alpher extrapolated from Humason's and Hubble's conclusion combined with Lemaitre's idea, published a description of their theory. They erroneously came to the conclusion 'that at one instant and place' there was a monumental explosion, later to be tabbed by a highly respected British astronomer Sir Fred Hoyle, as the "Big-Bang". Fred Hoyle named the concept in one of his weekly radio broadcasts. Hoyle's phrase was intended to discredit and be derogatory, but it was catchy and the phrase stuck. He favored a concept by Thomas Gold for "Continuous Creation of the Universe". Fred Hoyle, tried unsuccessfully to validate the continuous creation theory. In 1952, George Gamow, Russian born theoretical physicist at George Washington University in St Louis, Missouri, (after the big bang was widely accepted), Gamow supported the idea that universe expanded and contracted to a small size only to elastically rebound, (in his words, "und so weiter...."). Gamow never offered any explanation as to how this process originally started. Gold's concept speculated that as matter expanded "out of sight", or was consumed by stars (not being aware of black holes back then), other particles would periodically spontaneously appear elsewhere as a result of interaction of various energy waves. They speculated that the process was continuing in a steady state manner and termed it as a "steady state universe". In the debates that ensued, it was shown that there were insufficient energy waves available to support the Continuous Creation concept. (Why didn't they recognize that there was even a less possibility for such a humongous amount of energy and other matter coming from a zero dimension dot source of the BB?) The discovery of quasars in 1963 slammed the lid and ruled out Gold's theory. So the Continuous Creation Theory died. (Fred Hoyle's text book is an excellent resource for information on general astronomy and cosmology, and illustrates the wide spread influence of cosmology processes on related subjects. I think the NUT will have a wider influence on all thinking!).

An idea that received some credible recognition was the "Plasma Theory" proposed in the 1950's by Swedish astrophysicist Hannes Olaf Gost Alfven. However, his concept died from lack of adequate supporting analysis. (I could find none)

Several years before Hubble's and Humason's red shift confirming of Slipher's observation and subsequent claim for an (accelerating) exploding universe, a Russian physicist A. Friedman and a Belgian physicist G. Le Maitre independently proposed an origin for the universe as a hot dense state from which it has been expanding ever since. That idea was supposedly based on some of Albert Einstein's earlier work which others deducted to imply the same thing. (I have not found any records that show Einstein ever stated or implied that he accepted or supported the BB concept, although in an attempt to gain credibility for the BB, some claim he agreed). Friedman's and G. Le Maitre's ideas are said to be the true beginning of the big bang idea. It was publically introduced at least three years before Humason's and Hubble's confirming observations, but several years after Slipher's discovery of increase of distance and intergalactic red shifts, which was later "perceived as confirmation" of what I consider a flawed single point origin concept. That notion has hindered astronomers of the 1930's and for the next 75 years from thinking through the logic to my believed correct explanation of red shift and distance.

Over the eons since the arrival of mankind, there have been many ideas invented in an attempt to explain the origin of the universe. The New Universe Theory is the only known answer, that is not in violation of the Laws of Physics. A few historical 'Origins of our Universe' ideas are listed and compared in Figure 2.1.Table.

Some of Mankind's Ideas of the Universe's Origin			
Concept **Originator** **Era**	**Laws of Physics Compliant**	**Basis of Concept**	**Logic**
Giant & the Cow Bewildered Early Man ~20,000 to ~6,000 BC	Simple and Innocent. Laws not discovered	Primitive Imagination	None
Devine Beings; Chinese, Greek, Egypt, Italy, etc ~4500 BC to now	No science basis. Logic and Laws not known	Myth and Conjecture	None
Singularity Einstein, Friedman 1907 to 1922	None	Math and Speculation	None
Big Bang Georges Le Maitre 1927	Ignores Laws of Science	Interpolation of mis-interpreted Red-Shift data	Revers extrapolation of data
Continuous Creation Thomas Gold ~1948	Thought to comply	Assumed Cosmological Principle	Logic based
Plasma Hannes Olaf Gost Alfven ~1962	Not understood	Desperate for alternate to BB hypothesis	None known
Deflagration Mechanisms Bobby L McGehee ~2002 to 2004	Totally compliant	Red-Shift, relativity, and deductive reason	Logic + Laws & known facts

Figure 2.1. Table. Origins. From the dawn of intelligence and curiosity, the origin of the universe has been a subject of intrigue and speculation. A few man made concepts for the origin of the universe are examined "in the light" of the Laws of Physics. Only one complies.

The 'standard model' of 20th Century big bang idea is described with analysis and is presented by Steven Weinberg in his book "The First Three Minutes .....a modern view of the origin of the universe", (Recommended reading, see references), in which he refers to the "standard model". Weinberg presents his story, with the credible intention of making it comprehendible for all, including non-technical and semi-technical readers. Many more recent presentations of the BB idea are available by searching the web for the term 'Big Bang'. One of the more comprehensive which I recommend for reading is titled "Violence in the Cosmos" by Mike Guidry. It was, and hopefully can still be found at the internet web site: [http://csep10.phys.utk.edu/guidry/violence/violence-root.html]. I commend and hope the author Guidry does not remove or modify his impressive story at this site. All should have an opportunity to read his work, which is an excellent explanation of the (archaic?) 20th century 'beliefs'. To assure that I could refer to it in the future, I printed out a copy for my files.

All Big Bang synopses portray that in-and-at the initial explosion, a fountain of high energy photons, containing <u>all</u> of the matter of the universe, instantaneously erupted out of a single point. As that concept goes, the eruption happened at one instant, from one dimension-less point, but rapidly expanded to reduce the temperature from near infinity (gasp!) to the level where energy could be transformed into leptons (electrons and positrons) along with some low energy photons (electromagnetic light units). According to the BB model, after about one minute the matter further transformed into mostly protons and neutrons, plus some re-emerging photons. After a few minutes, free protons and neutrons became a mixture of nuclei of hydrogen, deuterium, tritium, and helium. The transition of matter from energy to mass particles is sometimes referred to as decoupling.

The "Standard Model" of the BB apparently was evolved from contributions by many astronomers and physicists of the 20th century. I have not been able to identify any specific individual to credit (or blame) as the developer or coordinator of the 'model design' effort. Apparently, the Physicist that put together the matter evolution part of the concept for the BB was George Gamow. Much of the in-depth analytical work by particle physicists are aptly described in Steven Weinberg's "The First Three Minutes" several book releases from1977 and through 1993. *Ironically, only by accepting some major, unreasonable assumptions, can the BB processes as explained seem reasonable.* Many of us, while reading his work, knew there had to be a better and more plausible answer; and as Weinberg himself said he suspected. Weinberg's 1993 book calculations on elementary particle masses and transformation energies are based on physics data which is from widely documented laboratory experimental data from the past 100+ years of laboratory experiments. Those data are from experiments by scientists, working with particle accelerators, such as CERN, and in other laboratories around the world, Weinberg's analysis goes on to "explain" the observed composition of the intergalactic concentrations of isotopes.

(*In modern test engineering, starting with an answer and some basic data, filling in the analysis between is called the 'dry-lab' process*).

These, now obsolete (since the revealing of the NUT) "standard models" are apparently available at many universities around the world. University of Washington at Seattle, Professor of Astronomy, Bruce Margon recently made an impressive TV Education Channel presentation (Ref 4.) on the BB (broadcast in early 2002). In the broadcast, Margon mentioned retiring. I say no way, Bruce, you have barely started your work, and we need your expertise. Bruce's presentation was "impressive"; Maybe after reading and studying this book, Bruce would be willing to make a presentation of this New Universe Theory through a similar TV educational show for the public.

The Big Bang? No way! Presentation of the New Universe Theory could include: Paraphrasing, updating, and embellishing Steven Weinberg's and Bruce Margon's comments,..... in place of saying "Sometime after the first few microseconds",..... It should be said ..... "It requires between 2 and 3 billion years for the proto-universe proto-matter to transform into the first observable mass objects and appear in the universe. It took a few billion years longer after the initial matter transfers for the particles started forming which later built our region of the universe. Still another 13 or so billion years transpired for the astronomical objects to evolve and grow into the complex objects, structures, and superstructures we observe today from our vantage point in time and space. The amazing thing is that more proto-matter is continuing to transform into more mass and energy, and <u>our universe is not exploding to its death; on the contrary, it is growing with vim and vigor</u>. It is growing at the speed of light and therefore is currently more than twice the size that is possible for us to observe. Proto-matter and its continuing transformations are explained in the following chapters.

The new Universe Theory story is so exciting it is difficult, if not impossible, to reveal this astounding phenomena in an orderly manner. But I will try. Read on!

Chapter 3 ... Red-Shift Mis-Interpretation

But what really is red-shift?

Red-Shift, is simply an example of the Doppler principle, which was defined in about 1850 by Christian Johann Doppler. The Doppler principle is the phenomena that modifies the observed frequency of a wave that emanates from a source that has velocity. In the case of a train passing by, the sound waves are stretched as the train passes and speeds away. Stretched sound waves are lower frequency and therefore the noise from the train is a lower pitch. Light from a source moving towards or away from the observer will have the light waves respectively compressed and shortened, or stretched and lengthened. Visible light consists of wave lengths in the range of 3150 Angstroms (A) up to 7700 A. (An Angstrom is a one-hundred-millionth of an inch) The shorter waves are perceived as blue and the longer waves as red. The objects moving away have the wave lengths of their light lengthened, shifting the observed light towards the red end of the visible spectrum. This is the so called, "Red-Shift".

Bobby McGehee

How could something so simple be misinterpreted by so many?

In 1912, the first astronomer to observe Doppler shift of light from space objects was Vesto Melvin Slipher at the Lowell Observatory in Flagstaff Arizona, who discovered light from the Andromeda nebula is shifted towards the blue (11 years later this nebula was re-discovered by Humason and Hubble to be a galaxy, similar to our Milky Way). The blue shift means our Milky Way Galaxy and the Andromeda galaxy are moving towards each other (at 10,000 miles per hour). (Appendix 2.0 includes more on Doppler and Red-Shift)

Early in the 20th century, it was discovered that the farther away an object is the greater is the red shift. Throughout the 20th century, this was considered as proof the universe is accelerating as well as expanding, and it also was claimed that everything originally came from a single point. This is what misled not only those saying it, but most other 'scientists' for the next eight or ten decades during several following generations. Law violations are an integral part of the big bang ~~theory~~ hypothesis.

The first and biggest violation by the big bang hypothesis is of the First Law of Thermodynamics which states "Matter cannot be created or destroyed." Another violation was of Newton's 1st law of motion; "A body in motion or at rest will not change from that rest or motion unless acted upon by an outside force". The big bang concept of accelerating expansion continues to be envisioned as real, but it has no continuing force to provide the initial or the 'observed' continuing acceleration.

What is going to change, is the red shift interpretation. The cosmological red-shift indicates higher <u>velocities</u> for objects farther away, <u>not acceleration.</u> The objects that are relatively close are moving away at velocities that

are slower than the objects that are farther away, and in all directions astronomers continue to observe red shifts in direct relation to distance. Measured red-shift (velocity) divided by a measured distance yields a specific number, which Edwin Hubble erroneously thought was a constant for all the universe. An erroneous hypothesis most astronomers continue to accept. From the on-set, astronomers have been uncomfortable with Hubble number diversity. Determining distance for red shifted objects would be simple if a single Hubble number applied in all directions and at all distances, as has been assumed, to date.

Hubble Numbers

Date Year	Number Km/sec/ MPS	Number Km/sec/ BLY	Observer Astronomer
1930	540	.017604	Humason, Hubble
1949	260	.008476	Baade
1955	180	.005868	Humason, et.al.
2000	117	.003814	Gegis, Mittapalli, Choi
1975	100	.003260	Van den Bergh
1979	100	.003260	De Vaucouleurs
1980	95	.003097	Aaronson
1983	82	.002673	Aaronson, Mould
1977	80	.002608	Tully, Fisher
1956	75	.002543	Santage
2002	74	.002445	Edward A Ajhar
1980	65	.002119	Mould, et.al.
1974	58	,001826	Santage, Tammann
1982	50	.001630	Santage, Tammann

Figure 3.1. Table of Hubble Numbers. Since the discovery of Hubble Numbers about eight decades ago, astronomers have measured them over a range of values. Fourteen typical values are listed in this table. The numbers represent the proportion of higher velocity for objects observed at greater distances. Velocity is conventionally expressed in kilometers per second and distance is clearly understood in "light years', but Hubble used 'MegaParsecs' and other

astronomers continued this convention for calculating these misunderstood numbers. (Units are described in appendix). More straight forward and meaningful units for large distances are 'Billion Light Years'. (Hubble number Data sources: Text; "Exploring the Dynamic Universe", by Theodore R Snow; NASA-APOD, and; Sky and Telescope magazine.)

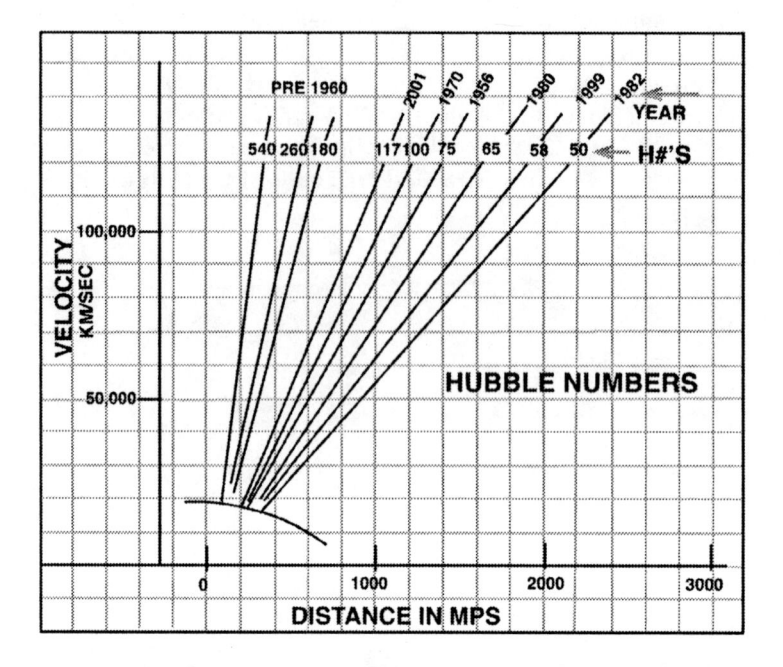

Figure 3.2. Hubble Numbers. Representative Hubble Numbers from the Figure 3.1 Table are illustrated in graph form to illustrate velocities verses distance. Astronomers measure the red-shift from a distant object and thereby calculate the velocity of the object recession. Distance is based on some other information sources. With the velocity and a Hubble Number (line), this graph allows distance determination for red shifted objects. It is understandable why astronomers have been trying to arrive at a single "standard" number and line.

The 'scatter band' for Hubble numbers, since they were invented more than 75 years ago, has been narrowing. Could this be because more Astronomers are now looking in the same direction when getting data for arriving at their Hubble number? Or are most now using the same distance candle? Or both? Or is it due to subjective peer pressure?

The scatter-band for Hubble number data has become smaller over recent decades. The NUT reveals why they are different in different directions.

Bobby McGehee

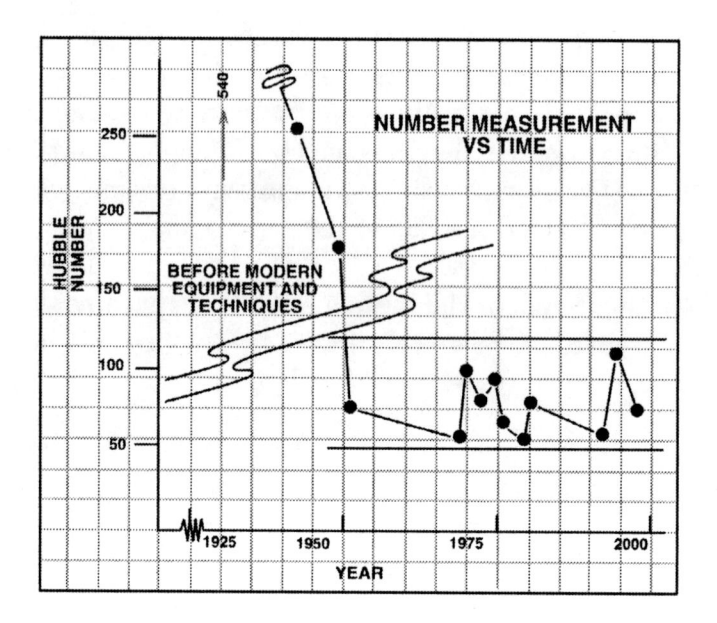

Figure 3.3. Hubble Numbers and passing of time. Graphically illustrating some typical Hubble Numbers compared to the year they were obtained shows how the measurements have varied over the past 75 years. To determine Hubble numbers, red-shift is measured, velocity is calculated, and the velocity is compared to the distance of another object in the vicinity whose distance is known by other measurement methods. Those known distance and velocity objects are said to be 'distance candles'. It is assumed that all of these measurements were made by competent Astronomers, and they simply measured their objects in different directions. There are several other factors that were not considered. Astronomers have been working very hard to standardize the Hubble Numbers into one value. Sub-consciously motivated to make data comply and support the BB hypothesis?

All Hubble (H) numbers are calculated from red shift measurements, which are valid indicators of specific light source velocities. But H numbers are also based on distance information based on various other "Accepted distance candles" which are then correlated with the specific red shift data? It is time to open our eyes and take another look at Red-Shift observations, especially those used to calculate Hubble numbers. My analogy for the New Universe Theory Red-Shift, which is consistent with the laws of Physics and clarifies how the red-shift indication of higher velocity objects, that are farther away, were easily misinterpreted as "proof" of acceleration:

When our son, (Bobby Jr.), was about 3 years old, he would come into our bedroom <u>early</u> in the morning, and while I was still sleeping, he would reach up, gently open my eyelid and say; "Open your eyes and look at me"!

The red-shift has been saying that to all of us for several decades. Following here is an eye opening analogy: Visualize a "down to earth" type occurrence; Assume that we are standing at the side of a rough, bumpy road, and along comes a light pick-up truck traveling at 50 miles per hour. The truck has its tail gate open and is carrying a partial load of potatoes. As the truck bounces down the road, and potatoes in the truck bed are tumbling around somewhat violently, a few of them are being regularly expelled out of the light truck onto the road. As the potatoes bounce and roll down the road, they are gradually slowing down and eventually stop. During the entire process, all of the potatoes are traveling away from our roadside observation position. <u>Decelerating</u> away from us as they individually give up their linear momentums to rotational momentums by way of friction, and through collisions with each other and the road. Meanwhile the truck continues down the road at 50 miles per hour, distributing potatoes.......... If we could from our roadside position, measure with a radar *gun, the velocities of the individual potatoes, we would observe the ones farthest from us (those closest to the truck) are moving away from us at a higher velocity than the ones closer to us. The potatoes farthest from us are receding from us at almost the 'speed of the light truck'; the ones closer to us are also traveling away from us, but at lower speeds. If we could only see individual potato velocity data, and we simply connected the dots, we might conclude that the potatoes are all accelerating away from us, at a rate approximately proportional to the distance from us, and <u>they appear to be accelerating</u> towards the light truck speed!. We know that is not the case, as the potatoes <u>are actually decelerating</u>, while continuing to move towards the light truck. I believe the same is true for the galaxies of the universe. They are decelerating at various rates, away from the central universe and towards the mass generating 'deflagration front', which is moving at light speed. Yet those farther from us, ... closer to the wave front have higher velocities. .... For 75+ years, no one has seen the potatoes! .... (Or was that a turnip truck?)*

Before we leave the potato analogy, let us consider one more fact. If we allow that a little bug is clinging on to one of the potatoes that is flying through the air between us and the truck: As the bug looks around at potatoes in both front and backward directions, his potato and all others appear to be, and in fact are, separating from all other potatoes. The farther away potatoes are separating at faster rates than those that are closer by, (in all directions). If there was another truck traveling in the opposite direction spilling potatoes, they would also be separating at another rate. If our pet bug was cunning, he could come up with a rate of separation compared to distance. Others might observe different directions and the "numbers" would be different. If yet another bug analyzed his data and found a variation in rates, they might think there was a peculiar influence which could be called inflation.

Moving on from the potato (or turnip) truck analogy.....

Sounds like double talk? Not at all! Hubble and many others have been had by the same trap. *I really like the statement by Carl Sagan: (ref 9) "There will come a time when our descendants will be amazed that we did not know things that are so plain to them".* Red shift only indicates velocities of individual light emitting objects. Acceleration is an erroneous interpretation from a false assumption that one can simply connect the dots. Actually, the objects all have rates of deceleration towards the deflagration front from which they precipitated and which is receding in all directions from the origin, also in all directions away from us, (relative to our location, wherever that may be). The observed red-shift measurements to date are for different <u>individual</u> objects at various specific velocities, the higher red-shifts (objects receding faster) are for objects that are closer to the deflagration front. By the Laws of Physics, objects near or far from us can <u>never increase</u> their individual speed outward. Only decrease their speed outward, and therefore are in a state of <u>deceleration</u>. Deceleration or acceleration could only be measured by taking several red shift measurements from the same object over a time period, but the deceleration is so slow, this compound process would take several hundred thousand years to obtain enough difference between two red-shift measurements for detection.

The central universe is slowly approaching a dynamic steady state, eventually with almost none of its objects receding from each other. Mutual gravitational attractions between the numerous objects results in coalescence concluding as rotations and vortexing of all objects of all sizes, up to and including, intergalactic structures and superstructures. As Allen Shapiro (father of modern thermodynamics) once said, "Vortexes never die, they just fade away". The central universe will have large dynamic rotations forever. The core of the central universe may eventually be swept by the arms of rotating superstructures. However, it is my belief that the core of the central universe will be a several-billion-light-year region, continuing to forever be void or at least rarefied of matter. (Rare, even by the standard of interstellar average densities) Not unlike the fig analogy described in Chapter 4.

Hubble number(s), contrary to belief for the last 80 years, are not rates of expansion. They are, ... relative-to-us, <u>the rate of measured differences of velocity</u> between distant objects in the universe, while those objects are traveling away and are <u>decelerating away from us</u>. In engineering terms, Hubble numbers are measures of velocity gradients measured across some parallel and some diverging flows.

In some directions, the red shift indicates we are moving away from the objects that were formed before our region of the universe, by the deflagration process. In all directions the objects are, moving away from us. We cannot easily distinguish between the ones which are older and the ones that are younger than our region. (<u>But we will in due time, ... keep reading</u>).

From the first red-shift measurements that led to calculations of Hubble numbers, Astronomers have been trying to "correct and eliminate" the variations for the values of the many Hubble numbers. My thoughts are; the probable answer is; "Most of them are already correct" as explained later in this work. The following was excerpted from a document downloaded from the APOD site at: <<u>http://antwrp.gsfc.nasa.gov/diamond_jubilee/d_1996/tvabout.html</u>> :

Long-long ago, in April, 1920, Harlow Shapley and Heber D Curtis debated "The scale of the Universe" in the main auditorium of Smithsonian's Natural History Museum in Washington, D.C. Again in April, 1996, Sydney van de Bergh and Gustav A Tammann debated the "Scale of the Universe".

Again in the main auditorium of the Smithsonian's Natural History Museum in Washington, D.C. This time the debate centered on controversies surrounding the determination of the Hubble "constant" which they thought was the expansion rate of the universe (So they said). This single number, by their interpretation, parameters the size and age of the Universe. Two large camps have emerged backing different values of Hubble's constant. (My belief is that the observations and numbers by the individual astronomers of both camps are valid. Both camps have erroneously assumed the Hubble number is a constant and that there is only one number. Perhaps when they have studied the New Universe Theory in this book, they will reach an amicable consensus. I believe they will conclude that all of their colleagues are also competent observers.)

It is now almost a decade later and there continues to be a lot of controversy. It is time to recognize that the Hubble number is not universally uniform and is not a constant. The numbers all have value, but are different in different directions and at different distances. There is a range of valid Hubble numbers. *However, in Slipher's, Hubble's, and Humason's honor, it is my suggestion and recommendation that the various values always be referred to as Hubble Numbers, (not constants and not law).* Data variations and differences are mostly thought to not be due to the level of competence, equipment sophistication, or skill of the measurers. <u>The differences are</u> believed to be real. <u>Facts.</u> <u>Not simply measurement errors.</u>

There are numerous articles by authors that subscribe to the Hubble expansion rate theory. For fun, I searched the internet for "Hubble Constant" and received 24,053 hits. Searching for "Universe Expansion Acceleration" yielded 10,406 hits. Some are interesting, most are humorous, although they are not intended that way. The writer's motivations are good as their intentions are of course to disseminate knowledge. Unfortunately it is almost all based on erroneous interpretations.

At the October 2001 Sun City West Astronomy club meeting, we were honored to have Professional Astronomer Margaret J. Geller, accompanied by Harvard Professor of Astronomy John Huchura. As the guest speaker, Ms. Geller made an informative presentation of the then present status of their mapping of the universe project. During the question and answer session, one of the members asked Ms. Geller, what was her favorite value for the Hubble Constant? Her curt reply, "I don't care as that doesn't affect our survey which is based on red-shift". She apparently recognizes Hubble numbers are not limited to one value. She promptly changed the subject and proceeded with her informative presentation. Ms. Geller is bright and perceptive. She readily recognized a hot potato!.....

My version of red-shift interpretation scuttles the BB and leaves us with the need for a feasible and valid theory. The "<u>New</u> <u>Universe</u> <u>Theory</u>" (NUT) is scientifically valid and fulfills that need. It is recognized that much work is yet to be done to refine and make the NUT more precise. The NUT follows scientific logic and also is consistent with the Laws of Physics. Three brief analogies are presented in Chapter 5 to provide a mind's eye view of the mechanisms of the New Universe Theory. A more direct and more comprehensive description is presented in Chapter 6.

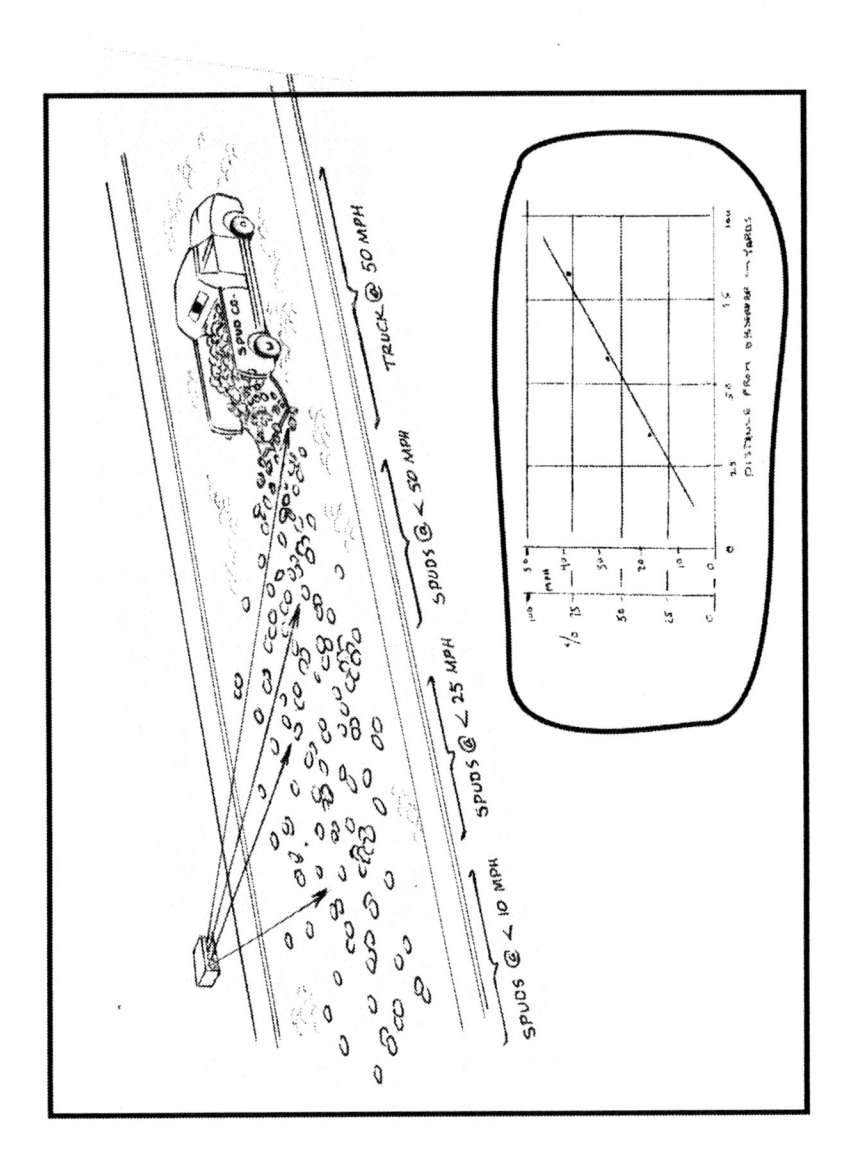

Figure 3.4. Spuds Decelerating. Speeds of individual spuds are measured with a Doppler device (radar) from roadside. Potatoes could be mis-interpreted as accelerating. They start decelerating th instant they fall from the truck, yet they continue rolling forward. The graph shows the spud speed (mph) is faster for potatoes that are greater distances (yards) from the radar monitor at roadside. It is just like a Hubble graph for galaxies' distances and red-shifts.

Chapter 4 ... Big Bang Theory

Big Bang and the Laws of Physics

Just because we don't immediately know the answer to a question about an observed phenomenon does not mean that a supernatural answer is acceptable. If it is outside the Laws of Physics, I believe we should <u>always</u> keep looking until an answer is found that is in full compliance. <u>There is always a scientific answer, we just have to find it!</u> The "old timers" (most are younger than I), had it right that the universe is expanding, but there is no evidence its galaxies are accelerating apart. This can clearly be seen as an astounding revelation as the New Universe Theory unfolds. Here are a few reasons why you should not believe the Big Bang (there are more, which will be discussed with the NUT).

Big Bang Claims :

The claim is:......... All matter, both mass and energy spouted in all directions and came from a single point with infinite density and temperature with no dimensions. Supposedly all matter not only came out of that single point, but also was expelled with very high velocity and unrelenting acceleration.

Violation: .... The First Law of thermodynamics: Matter can neither be created nor destroyed. Infinity is unachievable.

Comments: .... Mythical magic, super natural, and pseudo sciences do not comply with the laws of physics. Scientific thought is that all observations can be explainable with facts and logic.

The claim is: .... All energy and continuing expansion in the universe is a result of a single explosion.

Violation: .... Newton's Laws of Motion; An impulse produces an initial velocity, only. <u>Not</u> continuing acceleration.

Comments: .... All matter, objects and surrounding space not only is separating from the region of our galaxy, but also <u>appear</u> to be continually increasing their velocity, (accelerating), being propelled by an invisible continuing force. Yes, an impulse can produce momentum, imparting initial velocity to mass, but does not cause continually increasing velocity. The Law clearly states that continuing acceleration or deceleration requires a continuing force.

> **Analogy:** *Consider the discharging of a shotgun upwards. The pellets are launched initially at the 'muzzle velocity' by an explosive impulse. The pellets continually decelerate under the continuing force of gravity and aerodynamic drag. <u>The pellets never exceed their initial launch velocity, as there are no additional forces applied.</u> This would produce the same result if the shotgun were to be discharged in space from an orbiting satellite, without gravitational influence, but also in space there would be no aerodynamic drag. If the discharge were horizontal and perpendicular to the satellite's direction, the pellets would travel continuously along a velocity vector, at the*

satellite's velocity in one direction and at the muzzle velocity in the other direction; the pellets' resultant velocity would never change from their initial velocity until some thing or some other force interacted; After the initial impulse, and until interaction, there would not be any further acceleration or deceleration.

The claim is: .... there is an expansion constant. It is usually referred to as the Hubble number. All of the descriptions of the big bang claim there is a constant of previous and continuing acceleration.

Violation: .... Newton's Laws of Motion, (probably the best known); Force equals Mass times Acceleration. The BB supporters offer no explanation for this obvious violation. Acceleration, but no Force? Obviously a farce! Some have suggested <u>dark energy</u>?

Comments: .... When red-shifts were measured from distant galaxies, the farther galaxies are determined to be moving faster away from us than the closer ones. When these measurements of velocity and distance are plotted on a graph (speed units of Kilometers per second verses distance units of Mega-parsecs), the slope of the line is the apparent expansion rate; Later boldly claimed by Edwin Hubble as the "Hubble Law". The value of first measurements in the 1920's were calculated as 540, but many numbers were measured over the next several decades, and each decade produced other number values. They decreased to as low as 50 in the early 1980's and has since been calculated at anywhere from 50 to as much as 117 in 2001. Astronomers have had several conferences trying to decide which number is correct? Most now have 'standardized' their errors and have settled on an approximate number of 75. My opinion is that most of the numbers over the last 50 years are <u>all</u> valid!

The claim is: .... Inflation! Over the last decade the Hubble numbers have been determined <u>not</u> to be a constant, but they <u>appear</u> to be changing, usually increasing, with time. Some calculations show the variation in apparent acceleration to have been faster early on, then slowed, and now is increasing.

Violation: .... Newton's Law of Motion always has and always will apply. Not only does acceleration require force, but increasing accelerations would require additional or increasing forces.

Comments: ....Some 20th Century astronomers identified this apparent phenomena as inflation. But we are now in the 21st Century, and it is time to get real! The various Hubble numbers measured over the last several decades have been measured by competent astronomers using calibrated and reliable equipment. The numbers were simply measured at different distances and in different directions, and used different distance references.

> Rogier Windhorst, a bright young astronomer from Physics Department of Arizona State University was the guest speaker for the October 2001 meeting of the Astronomy Club of Sun City West. His subject was "Expanding Universe". His presentation was informative and indicative of the then current thinking of the most informed astronomy community. He and his presentation was appreciated and well received by an audience of well informed astronomy enthusiasts. To Rogier's credibility, he readily admitted that he did not understand how this phenomena could have occurred. After the meeting I told Rogier that I did not believe that inflation was real, nor was expansion, as defined by the Big Bang. I deduced that he does not fully accept inflation as real either, and I hinted at my Idea of a New Universe Theory; he momentarily perked up his ears, but time did not allow

us to pursue the subject further. I mention Rogier's presentation, as that was the impetus for me to get busy and document my thoughts which resulted in this work. I am indirectly indebted to him.

The Claim is: .... There is a Cosmological Principle. This "Principle" as claimed states that everywhere in the universe, all matter is distributed uniformly when viewed over large regions of space.

Violation: .... Fabrication. This is a broad and bold statement with no scientific foundation or purpose, only supposition. It just sounds good. The principal is not true even if the "large" region includes all of space.

Comment: .... This causes the expectation for, and feebly supports the concept of an exploding universe. There is no proof that our universe is exploding, and my contention is that our universe is not and never was exploding. Never the less, there is no known reason for this "principle", except to imply support for the Big Bang idea. It appears to be just excess baggage.

In my opinion, Violation of any (or just one) of the Laws of Physics is sufficient proof that the big bang has no credibility. Without valid explanations as to why the laws should be changed, the evidence is overwhelmingly adequate to drop the big bang concept, and find a better explanation. As suggested by Timothy Ferris, "Our descendants may look upon our theories,............, with scorn, bemusement, ridicule, or just hilarity,.........."
"Well said, Tim".

Now it is time for us to get real!

Grote Reber, famous Pioneer Radio Astronomer never-ever accepted the big bang. From this point forward, in

this book, occasionally BBB is used to relate to the term that Grote Reber used in a lecture, he titled "Big Bang is Bunk" (BBB.)

Read on.

NOTE: All matter in this book pertains to 'corporeal matter'. Corporeal matter includes all physical and tangible material, both mass and energy. Non-corporeal matter refers to word use such as; 'matter of fact' or, 'front matter in a book', or, 'his opinion doesn't matter'.

Chapter 5 ... Analogies of the NUT

What is this New Universe Theory? It is a consistent with the Laws of Physics concept that explains the production of the universe we observe and it is consistent with my new, interpretation of red-shift. Since this concept is totally consistent with the laws of physics, it thereby provides credible explanations of all observations of cosmological and astronomy observations. The concept is that the universe started from primordial matter (Stage I) and developed into what we can observe (defined as Stage III). It continues to be generated by three deflagration front reduction mechanisms, (described as Sub-Stages IIa, IIb, and IIc). Stage I proto-matter is being processed via the on-going Stage II reduction mechanisms into astronomy objects, which are being added continuously onto our Stage III observable universe.

According to the New Universe Theory, the universe did not come from an "in your face" big bang. All parts are decelerating away from us, therefore the mass generating wave must have come from somewhere behind us in space. It also came from somewhere before us in time, and generated our observable mass as it passed through this region. Future observable objects are being generated from this continuous on-going process that started at least a few billion years before the mass was generated that later evolved into our region of the universe. Our region includes our Milky Way galaxy, the local group, many other nearby super-clusters, including everything within a distance of about a billion light years. The mass generating deflagration front continues now, and the universe is currently an astounding 34.4 Billion light year diameter sphere which continues growing by 20 light years every decade.

These ages and the time periods are substantiated later in this document. Estimates are based upon growth rates deducted from extrapolation of Hubble numbers, which represent something different, and much more than their discoverers ever thought.

The New Universe Theory portrays that the distant objects which we are observing, are younger and are moving away from us. They are progressing outward and away, and at the same time they are de̱celerating to lower velocities, towards the now 'farther out' light speed transcending front. All the distant astronomical objects we observe were generated from that wall of transformation, just as we were several billion years earlier. But we have decelerated for a longer time and are now moving at a lower relative velocity. The deflagration wave (NUT shell) traversed the location of the far away objects in the direction of the Abell clusters long ago, but the shell simply traversed our location between .75 and 2.5 + billion years earlier than the Abell region (the Abell clusters are spread over a space volume between .75 and 2.5 + billion light years distance.) We simply started decelerating earlier than some galaxies, and later than those in the other direction. Those galaxies which were generated before us are moving outward at a slower speed than we, yet we are also decelerating. The distance between us and those slower moving galaxies is also increasing because we are moving faster. This gives the illusion of them accelerating in the opposite direction. Our having decelerated for a longer time than Abell clusters, we are at a lower velocity while traveling in the same direction. Relatively, we are separating from these younger objects as they are closer to the front and are therefore at higher velocity.

This all initially appears complex and at first the explanation is hard to follow, however it will become clear as this New Universe Theory description unfolds in the follow-on text. The preceding old 20th century exploding view was concluded from an optical illusion.

The 'old timer' astronomers (almost all of the living old timer astronomers are younger than I) were right in the presumption that the universe is expanding, but only in the sense that it is growing and getting larger. The newer and older objects are all decelerating outwards; <u>Not accelerating.</u>

Three analogies follow which help to convey the image of 'how the universe is growing'.

<u>*Analogy #1.*</u> *An interesting biological analogy that helps in visualizing the growth of the universe is provided by nature. The New Universe Theory growth of the universe can be compared to a fig by the similar manner in which they both grow. If you cut a fig in half, it is clear that the fig grew from the outside skin towards the center. Actually, the skin grows outward, like the front, away from the starting point, 'generating' and leaving behind the produced fig-mass. (I grew tasty king figs like these in my Seattle back yard.) Obviously, the center has no where in which to grow, so we know the outer portions are growing outward! The fig of course, gets larger in diameter as growth continues. The peripheral skin surface area adds internal volume as it continues to progress outward as the fig grows, just as does the deflagration front at the periphery of the universe.*

The growth of both of these two "skins" result in more internal volume that accommodates more interior mass additions, being produced from their skins. An interesting observation is that the central region includes a void region and is also becoming less crowded as the fig grows. In both cases, the universe, and the fig, new mass is being added to the interior as the periphery advances outward.

NEW UNIVERSE THEORY WITH THE LAWS OF PHYSICS

Figure 5.1. Fig Analogy. These are fresh Washington state figs. The photograph of one cut in half shows how the interior growth is from the skin, but towards the center at a slower rate than the skin as it grows outward to increase the fig's volume. A biological analogy that evolved into a pattern of growth similar to the growth of the universe provides us with a physical and visible example of the New Universe Theory.

Analogy #2: *To visualize the combustion propagation analogy, consider an August lawn of dry-brown grass and a careless passerby flips a burning cigarette butt onto the lawn; the grass ignites and the flame progresses in all directions. As the flame proceeds it burns and propagates (deflagrates) in a circle, and leaves behind combustion product ashes. These ashes are made up of the same material that was there prior to the passing of the deflagration front, it simply remains there in a different form.*

Analogy #3: *This analogy starts with a large combustible gaseous mixture that is made up of vaporized fuel and an oxidizer in fuel rich but near stoichiometric proportions. When we have a spontaneous ignition within the volume of gas, this time the deflagration is in all three dimensions and rapidly propagates outward in all directions as a flame-front that travels near, if not at the speed of sound. The products of combustion, both energy, chemical compounds, smoke particles and other combustion products, are expelled to the rear of the propagating front. However, these "ashes" travel in the same direction as the front, initially at the deflagration front speed, but their forward speed slows due to vortexing and mixing. The outward flow of the combustion products are slowing behind/within the propagating deflagrated zone. The front proceeds through, and farther into the combustible mixture, (the product material is the same quantity, but is in a different form) the products from the front are of course confined within the volume of the combusted sphere. As the products of combustion mix, they decelerate, through minute transverse motions in vortices and also the new compounds coalesce into smoke fragments and the fragments clump. Behind the combustion front, is a sphere of clumping and mixing of combustion products, which is analogous to the central universe. The volume of combustible mixture outside the deflagration front is analogous to proto-matter and proto-space in our proto-universe. The products of combustion follow in the same direction as the propagating spherical front, and continue to decelerate towards zero linear velocity. In both the analogy and the Universe, deceleration is as a result of rotational mixing and vortexes which are consuming the linear momentums but are continuing outward, at lower velocities within the sphere of 'generated' products.*

The New Universe Theory includes the 'matter transformation' powered wave, propagating at the speed of light; it can be visualized by relating to the analogies of the described chemical processes called combustion deflagration that also propagates outward in all directions.

The Universe's generating mechanism (deflagration front) travels at the speed of light and the transformation of proto-matter is from positronium mass particles, to energy, and then into other matter. These 'other matter' mass particles, then coalesce into clumps, to form all sizes of objects, from nuclides, to stars; globular clusters; galaxies; quasars; super-galaxies; structures; and then superstructures. This process is continuing to propagate, long after, the same process started occurring in the central part of the observable universe. Since we are not located at the starting location of the universe, some of the observable objects were generated before us and some were generated after us. Yet, all are separating from us, and each other, with time.

All of the transformation of matter in the universe involves conversion of linear velocities and momentum to rotational velocities and angular momentums. (There are no mythical or mysterious phenomena available to act as accelerating or decelerating forces applied to the universe's objects.) These linear to rotational momentum transfers are physical processes and occur consistent with the **Laws of Physics**. The even more distant proto-matter and proto-space volume beyond the universe will continue to be processed and added onto the growing universe's volume and in the future will become more future objects and space. These new products then evolve and mature into the continuously growing numbers of stars, galaxies, quasars, and other objects in the "Observable Universe". The New Universe Theory portrays that the transformation of matter is from annihilation photons, through an understood but complex process, into the visible objects.

Even in the BB hypothetical concept of an instantaneous event, the so called 'standard model' required time for the products to develop?! ..... Calculated therein was supposedly a one shot transformation that required about a billion years, more or less, to produce the first stars. There are many major significant differences in the matter evolution between the hypothesized one time event BB and the scientifically feasible and continuing New Universe Theory. The BB model idea was that it occurred instantaneously from an impossibly infinite density and temperature (whew!), and then, over time, matter was said to be disbursed into the present universe to today's volume and density The BB high temperatures were said to have been necessary to give the particles adequate velocities and momentum (force) to cause fusions between colliding particles, and the fusions had to occur before these particles moved apart in their divergent paths. The hypothesized divergent flow occurred rapidly.

With the New Universe Theory, the transitional stage is continuing and can be understood as a peripheral shell with thickness, continuously progressing, and traveling outward at near the speed of light. Fusion force is from velocity enhanced gravity (explained later). The development **Stages** are in three concentric bands, and are moving outward and forward with, but behind the annihilations that are leading the deflagration front. This three layer NUT shell has a total thickness of about two and a-half billion light years (the thickness of the first two layers is measured in centimeters). Consistent with the "continuity" Laws of Physics, matter density in the universe (behind the Front) is identical to the proto-matter density beyond the "front".

The "transition zone" is defined as **Stage II**. It extends over the distance from the Stage III observable central universe to the **Stage I** proto-space and proto-matter interface, and includes the annihilation 'chain reaction' wave. This zone extends all the way through the initial clumping matter region". By definition, the **Stage II** region overlaps the outer reaches of the interior **Stage III** observable central universe. Stating again, the transition zone follows immediately behind the speed of light traveling annihilation wave.

Within the transition zone, there are decelerations of the precipitated mass (primarily protons and neutrons). Forming rapidly through fusions all nuclides and isotopes instantly acquire the combined mass characteristics of gravity and inertia. Interactions of these particles with extremely high velocity-enhanced gravitational forces, results in transfers of linear momentums to angular and rotational momentums. Farther forward at the annihilation front, is where matter and anti-matter (positive and negative) Leptons (electrons and positrons) are combining and converting into the advancing wall of photons (electromagnetic energy). Within the transition, photon energy concentrations convert back to mass, primarily Protons and Neutrons (Hadrons), possibly via quarks and other sub-Hadron size particles. Then velocity-enhanced-gravity forces causes coalescence producing fusions. Nuclides of several light isotopes are formed. In the subsequent clumping zone, both small and large objects interact and develop more spinning and more mutual orbiting, resulting in linear decelerations, via linear momentums converting to angular momentums consistent with the Laws of Physics.

As a result of orbiting and revolving, linear decelerations continue throughout the transition zone; this becomes more significant as larger vortices form, interact, and continue to grow. Linear momentums are transferred into angular (rotation) momentums. Many compound orbiting particles and objects coalesce. Over many millions of light years, Galactic assemblies of vortexes form and grow, and linear decelerations continue as linear momentums are converted to angular momentums. There are no reasons to expect the various directional Hubble (H) Numbers are common or constant, as they are related to continuously varying deceleration rates. The H-numbers, although not constant, later stabilize for about the inner (older) one-third of the **Stage III** observed universe. This implies that many of the Hubble numbers, as indicated by past measurements are possibly valid in the region where measured. The H-numbers only apply over the limited distances that correspond to less than the full extrapolation. Reference distance candles are not yet available at the great distances beyond those corresponding to about 1/3 light speed, (over ~100,000 Km per second). The numbers are valid only over specific directions as well as limited distances, because they are measured from our off-center location, not the center of the universe. These considerations explain much of the 'wild range' of original numbers obtained by Humason, Hubble, and other early on astronomers. They were viewing objects only in our Local Universe, in various different directions, as well as in a different directions from their reference distance candles.

This New Universe Theory is graphically illustrated in the beginning of Chapter 6, as Figure 6.1. (The figure is busy, but remember, it represents a huge complex process). Descriptions follow for our Universes' growth, **Stages; I, II,** and **III**. Matter Reduction Mechanisms are described as **Sub-Stages IIA, IIB,** and **IIC**. These are more understandable as they are more comprehensively explained as you read a little farther into Chapter 6.

Recent red-shift sky survey data have been acquired from several sky surveys. And now, a publicly accessible data base is being established which will list the data from two of the more extensive surveys. (I do not know the details as to whom is responsible for achieving this commendable project, but thanks, from all of us). These two sets of data are from the "2 degree Field" (2dF) and the "Sloan Digital Sky Survey" (SDSS). We anxiously await and anticipate when these data will be accessible (maybe in late 2004, even before this book is published?), for all of us to review and analyze from various perspectives. We can then review red-shifts in the same vicinity regions with (other) substantiated and verified distance yardsticks. A three dimensional map of objects by distance based on other (non-red-shift) distance measurement methods will reveal untold phenomena. These data, when correlated with other distance candles can then give us a map of velocity distribution within the larger deceleration fields. The need for such analyses becomes more apparent after reviewing the following chapters.

Substantiation of the New Universe Theory (NUT) has been accomplished many times by observations throughout all of astronomy. All astronomical observations known to me fit nicely with the NUT in compliance with the universal Laws of Physics. The observations, when related to the BB, have to be explained with deviations from the laws of physics, or by saying it was something we don't yet understand. NUT revisions and updates will be achieved, after the anxiously anticipated comments and reviews are received from audits by critically scrutinizing astronomers (amateur and professional), physicists, and other interested readers. When recent 2dF and SDSS data are regionally mapped and correlated with specific and directly applicable distance candles, H-numbers over incremental distance and regions can finally be cataloged and available to describe the three dimensional sky.

Chapter 6 ... New Universe Theory

The New Universe Theory develops <u>in three progressing stages.</u>

Stage I: **Primordial Matter (Proto-Universe)**
Proto-Matter and Proto-Space, is an endless monolithic hexahedron-lattice-crystalline-structure of positroniums stabilized by electrostatic forces from uniform spacing (~12.5 cm) and parallel axes of self synchronizing rotations.

Stage II: **Reduction Mechanisms (Deflagration)**
<u>Sub-Stage IIA.</u> Annihilation Zone (mass to photons)

<u>Sub-Stage IIB.</u> Photons Transform to elementary and sub-elementary particles

<u>Sub-Stage IIC.</u> Clumping of Mass (elementary particles transform into larger objects through velocity-enhanced gravitational force nuclide fusions, nuclide coalescing, clumping, and later, star ignitions)

Stage III: **Observable Universe**
Potentially Observable, with current or future technology, containing both light emitting and non-light-emitting objects, plus background radiation.

The preceding outline identifies the basic components of the New Universe Theory. The following graph; Figure 6.1, "The deflagration Wave", illustrates the concept. I presented this to two engineers, their comments were; "Wow!, what a busy chart!". Yes, that is required to show this complex system in one graph. They agreed.

The next figure is also the first sketch I made of my initial NUT concept. When taking on the challenge of analyzing 'development of the universe', I considered the known observational facts first discovered by V M Slipher. He found from his red-shift measurements; the farther away any distant galaxy, the higher its velocity of recession from our vantage. Red-shift is a very reliable way to measure velocity, but it can not detect acceleration or deceleration. **Slipher's observation, is undoubtedly the most significant contribution to astronomy since Galileo**. This knowledge and the Laws of Physics, scientifically and logically, appear contradictory. But, there is no acceleration as Hubble exclaimed. Since there are no continuing forces, there could not be continuing acceleration. However, there could be deceleration, as velocity can be slowed by phenomena other than force. This reveals that primordial matter of the universe was/is antecedent. It converts to photons and then to mass objects as it condenses from light speed. The mass particles initial linear velocities then decelerate from light speed to ever slowing speeds by entropy. Neutrons coalesce and decay while vortexing and further coalescing into rotating star systems, galaxies, super galaxies. No mysterious forces or 'dark energy' is required as envisioned by Big Bang advocates. Figure 6.1 illustrates the envisioned 'deflagration front' stages. Stages are defined which include the processes. Estimates of the distances over which these processes occur are presented later.

NEW UNIVERSE THEORY WITH THE LAWS OF PHYSICS

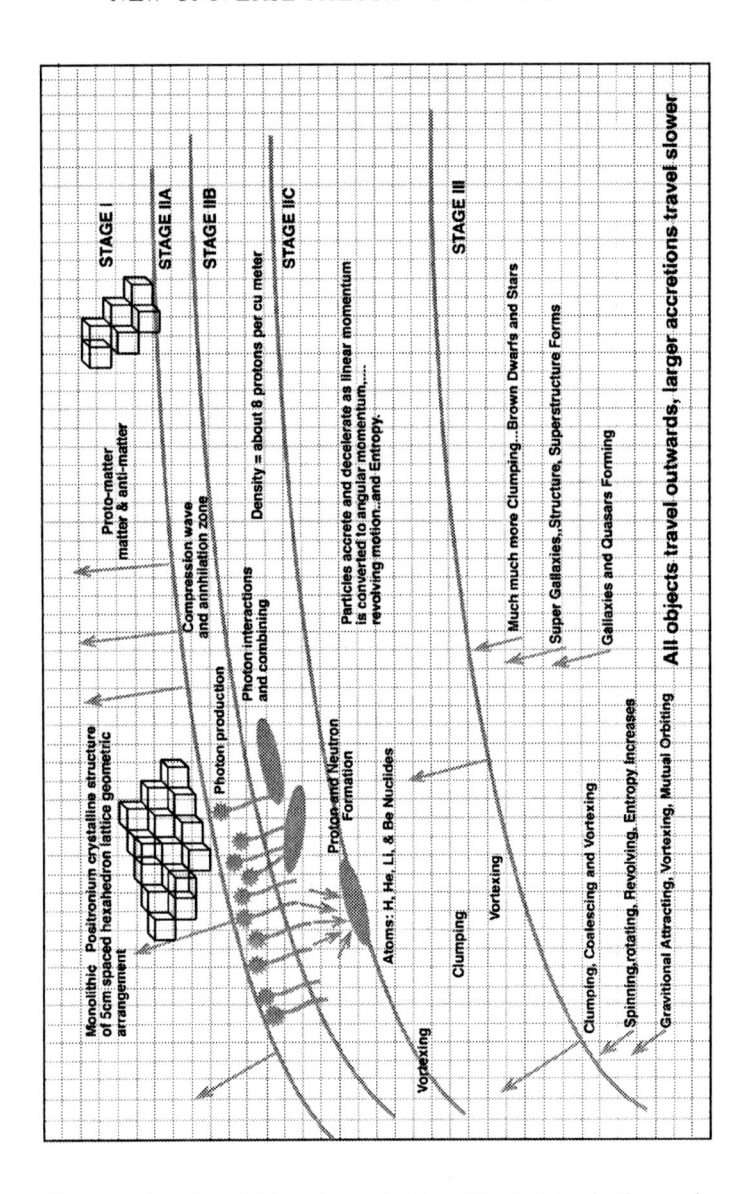

Figure 6.1. The "DEFLAGRATION WAVE" This figure is necessarily "busy" as it summarizes the New Universe Theory, revealing the largest of all processes. The universe is a three stage production: Stage I is the structure, space, and material from which Stage II generation and reduction mechanisms transform Stage I primordial matter into the ever increasing content and size of the dynamic universe. This observable portion is known as our universe, and is identified as Stage III. All three Stages are encompassed by the New Universe Theory.

61

Chapter 6-I ... Stage I
The Proto Universe
Primordial Matter ... Proto-space and Proto-matter

Definitions References: 27, 46, 47 by Webster's Encyclopedic Dictionary;

Primordial Matter *"Pertaining to the very beginning, the elementary stage".*

Matter *is "the substance or substances from which any object is made". Corporeal matter is made of physical and tangible material (mass and energy).*

Matter *includes all basic constituents, all of which are made up of either or both mass and energy. Both mass and energy have many forms. These constituents can be converted, or transposed to other forms; all mass, to all energy, or to combinations of the two.*

The Law of Physics that governs the transfer of these constituents is the famous relationship discovered by Albert Einstein. $E = mc^2$. The conversion factor between **m**ass and Energy is the square of the speed of light (c^2). Stage I of the

New Universe Theory is my theorized form of primordial matter, which is an assembly that is a single-huge monolithic crystalline structure that forms the entire Proto Universe. We have no way to estimate the size of the primordial proto-universe. Deep inside this primordial volume is a growing NUT shell that contains our entire 'expanding' and growing universe.

Understanding and arriving at feasible and explanations plausible to me including descriptions of primordial matter, requires a combination of both deductive reasoning and intuitive thought, and includes only substances proven to exist. Most astronomers and cosmologists now have what is thought to be a pretty good general understanding of the Stage III observable universe. Some amateur astronomers have a good awareness of Stage III contents. There is still, and always will be, a lot more that we don't know than there is that we do know. Now, The New Universe Theory, helps us to understand some knowledge we already have and exposes tremendous amounts of knowledge we have yet to explore and learn. With the proper interpretation of the red-shift, and with the remapping of inter-galactic velocity (red-shift, vs distance correlated) distribution, many revelations are pending. (This statement will be better understood after reading Chapter 10 and pondering Figure 10.2)

By observing the STAGE III objects and dynamics, recognizing the New Universe Theory transition stages, and projecting forward through these Stage II reduction mechanisms, we then arrive at primordial matter. The only acceptable and feasible answer to 'what is primordial matter' is through application of solutions and processes that comply throughout with the **Laws of Physics**. Primordial matter could be much more complex than the configuration I have defined and described in this work (more complex models were considered). However, the configuration described here is feasible, and probably the simplest. As found in nature and engineering, the simplest solution is the answer that almost always provides the best answer to a problem. Proto-Universe solution for the New Universe Theory, as described, is my thought as the most likely to be "the real thing". Much refinement is of course yet to come. It will be interesting to see what others might consider as alternatives.

Stage I, the proto-universe, contained all of the primordial space and material that, through the reduction mechanism processes of Stage II, is making up all of the matter in the observable universe, known as Stage III. Our observable universe contains a tremendously large, mind boggling, quantity of matter. Matter, <u>by definition</u>, always includes both mass <u>and</u> energy.

Most people are dazzled by big numbers. Especially astronomers and cosmologists. It is a good thing they like this, because, to coin a phrase, "........the numbers in cosmology are astronomical"............ Assume for the moment that the observable universe is a sphere 30 billion light years in diameter, then the universe's volume calculates to be 14,137 billion cubic light years. (A light year is equal to 5.9 trillion miles, and therefore each cubic light year contains 205.37 trillion cubic miles. Another interesting number is that the Schwarzschild Radius of the entire universe calculates to be 59 trillion-trillion miles (5.9 X 10^{25}), (Ref # 2). Verifying this value is mathematically tedious, requires use of vector analysis and differential equations, and is only an exercise in futility; just accept that whatever the specific number is, it would be the radius of the event horizon for a very large black hole. Also, as the universe continues to grow, that theoretical number will continue to increase. Contrary to the universe's big crunch cyclic speculators, this could never happen as long as the primordial matter supply lasts, as all linear momentums are outward, away from the center. The New Universe Theory concept shows that all astronomical objects, (and *that means everything) big and small, originated from the combining of large quantities, from an apparently limitless supply, of paired smallest <u>known</u> elementary unit electrically charged mass particles. If however, the supply is someday depleted, the universe's gravity and the continued deterioration of linear momentum, could maybe result in a big crunch, as envisioned by George Gamow. If this happens, the part about a rebound and the cyclic universe idea can never occur, because if contraction were to reduce the universe volume to its Schwarzschild Radius, it will never be able to re-emerge. (it is interesting to note that according to the BB scenario, all of the universe was once inside a single dot, much smaller than its Schwarzschild radius!; The definition of Schwarzschild radius is the size for an object at which the particles within it are so close together that the gravitational force, which by the laws of physics, is so strong that the particle and electromagnetic radiation would have to exceed the speed of light to escape. Laws of physics limits the speed of all matter (mass and energy) to the speed of light. The universe's Schwarzschild radius is 10^{26} Kilometers as calculated by Astronomy Text book author Theodore P. Snow (Ref # 2).*

Leptons

Particle	Rest energy / mass [1]	Charge
electron [3] (a.k.a. as negatrons)	511,000 electron volts (eV)	- 1.0
positron [3] (a.k.a. electron anti-matter)	511,000 eV	+ 1.0
Neutrino A [2]	0.0	0
Neutrino B [2]	.0000022 eV	
Neutrino C [2]	0.0	0
Muon	----	----
Gluon	?	?

Hadrons

Particle	Rest energy/mass [1]	Charge
Proton [3]	938,300,000 eV	+ 1.0
Neutron	939,600,000 eV	0.0

Quarks

Particle	Rest energy/mass [1]	Charge
u (up)	5,000,000	+2/3
d (down)	?	- 1/3
c (charmed)	?	+ 2/3
b (bottom)	?	- 1/3
t (top)	175,000	+ 2/3
s (strange)	?	- 1/3

[1] ..... Rest energy is the energy that will be released when a particle's mass is totally converted to energy.
[2] .... Recent (2002) research indicates there are three kinds of neutrinos, and some may have a rest mass.
[3] .... Strongly interacting particles.

Figure 6.2. Elementary Particles. Table of some elementary particles. Protons and Neutrons are theorized to be made up of still more elemental quarks. This New Universe Theory concept presumes that mass is precipitated from Photons into Hadrons, in accordance with Einstein's famous and proven mass-to-energy relationship, $E=Mc^2$. The process can go in either direction, energy to mass, and mass to energy. What transpires in and during this process is not totally understood, but never fear, particle physicists are busy unraveling the details. At this time, The New Universe Theory disregards the intermediate possible 'quark' steps and considers only the larger steps: 1. The transition from the Leptons (electrons and positrons) to photon energy gamma rays, and; 2. The subsequent precipitation of gamma ray photons to Hadrons (neutrons and protons). The table list includes some known, plus some theorized 'elementary' particles, to give some insight into the possible transition steps and processes which particle physicists have yet to answer. Elementary particles' rest energy or rest mass are sometimes expressed simply in energy units, as in the table.

We know from the First Law of Thermodynamics that matter can neither be created nor destroyed. No one disputes that the universe has had a finite life span, yet logic and the laws of physics tells us matter had to have pre-existed the universe. We also know that matter includes both mass and energy. The Continuity Laws require that the matter and space which enters the universe had, and will continue to have, the same quantity of matter and space per unit volume after arriving in the universe. Proto-matter and proto-space, the source of the universe, is being transformed through an orderly and continuous phenomena and this growth is through the entire outer perimeter of our universe. Actually, the perimeter of the universe is traversing through the primordial matter and as a result, the universe is growing in size in all directions in proportion to the increase in matter and space. The Primordial Matter (proto-matter and proto-space) is described and defined as **Stage I**. Particles and anti-particles are, and must be paired, poised for contact, but held in stable separation by the centrifugal forces from the rotation of the components. Positroniums (first discovered about 1945, references 27, 46, 47) are rotating couples, each made up of an electron and an anti-electron. (*Particle physicists sometimes refer to these electron particles as negatrons and positrons, respectively*) These pairs of particles are in a mutual orbit. In Physics of 'Statics and Dynamics', two non-aligned equal forces acting in opposite directions are defined as a 'couple'.

Einstein's 1905 developed relationship between mass and energy, comes into play in the concept of the New Universe Theory, which could never have been developed without Einstein's discoveries. He was first to demonstrates that matter can be transformed from mass to energy and back to mass without violating any of the conservation laws. The only conceivable source for the mind-boggling quantities of energy needed for the synthesizing to the mass of the universe comes through annihilations of mass and anti-mass particles which are stored in positroniums. Annihilation is the process of transferring mass to energy. The components are held together by mutual affinity between the positronium constituents from opposite electrical charges plus there are minuscule, but not insignificant, mutual gravitational attractions. The separating force that resists and balances the mutual attractions to maintain their separation is the centrifugal force which is a function of their rotational speed, their individual masses, and the distance of separation. As long as the forces stay balanced, there will be no contact, and therefore no annihilations. The balance is tenuous and is upset only when there are intruding forces. Positroniums have been observed in particle physics laboratories, but here on earth, and maybe in all of the **Stage III** universe, they have a short life (about one, ten millionth of a second) before their components' spontaneous mutual annihilation. The **Stage III** observable universe, in which we live, is permeated everywhere with random radiation; electromagnetic, and gravity waves from nearby random moving molecules, cosmic rays, and neutrinos, any of which can, and do upset the positroniums tenuous balance. Proto-space however is totally isolated from the universe and its randomness; this total isolation of the proto-universe provides the stable environment. In addition, there must be no temperature (temperature is a measure of the mean momentums of a substance's particles), because the pairs of electrons and positrons (positroniums) must also be locked in position both within the individual positronium and also relative to their adjacent neighbor pairs, otherwise contact would occur, resulting in annihilation.

Defined and described is this feasible concept for proto-space and proto-matter; density, structure, and material. This New Universe Theory concept begins with proto-matter and proto-space as monolithic positronium crystal that is in a hexahedron lattice arrangement, providing the stable environment of the proto-universe. When the positron and electron rotations and their positions are perturbed and their orbits deflected by intruding electromagnetic photons, the imbalances cause and perpetuate the cascading of annihilations which produce the continuing avalanche of high energy photons that subsequently combine to produce the universe's matter. All in full compliance with the stated and restated necessity for respect for the **Laws of Physics.** In particular the Laws for Conservation of Matter, and the Laws of Continuity; the matter density beyond and outside the deflagration front must be, and is, equal to the universe's average.

Dr. Sherman Eager, Professor and head of the Oklahoma State Physics department was teaching graduate courses in Electricity and Magnetism and when discussing annihilations; He often re-emphasized the fact that matter cannot be destroyed nor created, only converted to other forms. He also smiled and wryly said what I have found to be true; "If we are not careful about how we phrase our statements on this matter, we might be misquoted." I mention him because his frequent advice that we should always be skeptical and assure that all phenomena should be scrutinized for compliance with the Laws of Physics. He is probably why, for the last 50 years, I have tried to rationalize the BB.

At this time, and probably for several decades to come, the density of the universe's matter is not accurately known. (Not even close to accurately). It has been estimated by several astronomers, and their values vary widely, from an average of 12 protons per cubic meter, down to as low as one proton for every 16 cubic meters. For lack of a better number, I have assumed a convenient number within their range. My assumed average density of the universe is eight (8) atoms of hydrogen (protons) per cubic meter. The number of positroniums to provide this equivalent mass per unit volume is approximately eight thousand per cubic meter. The deflagration process functionality is not affected by density, therefore the choice of a convenient and feasible assumption of eight (8) positroniums per cubic liter is justifiable and is used for the calculations. With this density, the center of each positronium is perpendicularly separated from all adjacent positroniums by 12.5 centimeters, in all three spatial dimensions. The positroniums are separated on their common axes of rotation and the axes are separated by the same dimension. This is possible only if the rotations are in synchronization. Positroniums and lattice as shown in the following figures are not to scale. The lattice segments are 12.5 centimeters on all sides and the electrons and positrons are each only about 10^{-12} centimeters in diameter, (which is .0000000000001 of a centimeter, or one ten-trillionth of one centimeter).

Electrons and positrons are the smallest unit charged particles and are thought to be electrically charged clouds of matter. They each have one unit of charge, the electron has one negative unit and the positron has one positive unit. Thus the positronium is a dipole rotating couple, containing one electron and one positron. Their rotational centrifugal forces keep these electrical matter units separated, and their electrical and gravitational forces keep them from flying apart.

NEW UNIVERSE THEORY WITH THE LAWS OF PHYSICS

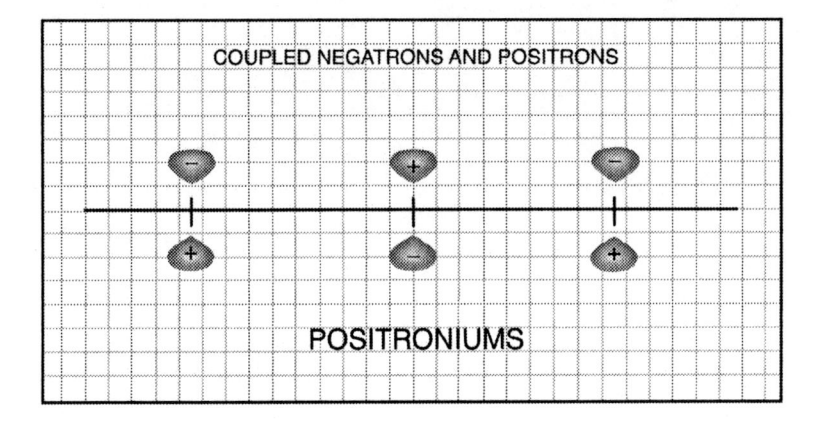

[Figure: Diagram labeled "COUPLED NEGATRONS AND POSITRONS" showing three positronium pairs arranged along a horizontal axis, with − and + charges above and + and − charges below, labeled "POSITRONIUMS"]

<u>**Figure 6.3. Positroniums.**</u> This figure shows three positronium pairs. Each positron and electron pair rotate about a common axis that passes through each of these couple's center of gravity. The continuous array on the axis are held uniformly separated by the electrical attraction and repulsion from the adjacent couple in both directions. (In physics of statics and dynamics, two misaligned equal forces acting in opposite directions result in what is known as a 'couple'.) In the universe the life span of a positronium couple is only about one-ten-millionth of a second. In the proto-universe at absolute zero temperature, would live forever if not disturbed.

In primordial matter (proto-matter), positroniums are rotating with their centers of rotation fixed in place. The positions are arranged in a stable hexahedron crystalline arrangement. Adjacent positronium centers of rotation are separated by 12.5 centimeters as are the axes of rotation, providing an overall density that corresponds to the average observable universe density. (The crystalline lattice figure shows only a typical few of the positroniums, however, one positronium actually resides at every vertex of the hexahedron lattice structure.) It must be understood, that not only is the universe's true average density not known, but most cosmologists and astronomers have the opinion from observations of galactic and intergalactic orbital paths and their relative velocities, that from gravitational observation, there are several times more mass in the universe than has been otherwise observed (maybe 10 or more times). Assuming this as valid, the continuity laws up the requirement for primordial matter density by the same factor, reducing the positronium separation dimensions accordingly. It also should be noted that whatever the correct true density is eventually concluded to be, the transformation and reduction mechanism processes do not change, just that the density is equal on both sides of the Stage II reduction mechanisms.

Why Hexahedrons? Why not Tetrahedrons, Octahedrons, Dodecahedrons, etc.,?

The rotational speeds of all positroniums must be equal and rotating in synchronization to be continuously electrostatically held in fixed positions. Adjacent positronium pairs must also be oriented so their polarities are oppositely aligned with all of their next door neighbors. The alternating pattern continues throughout the proto-matter array.

Geometric patterns other than hexahedrons can provide uniform patterns of positronium arrangement, but only hexahedron lattice provides uniform and equal spacing with adjacent lattice vertexes in all six directions. The only known lattice arrangement which can accommodate all of the stability requirements, is hexahedron. Therefore, this is my assumption as to the crystalline lattice arrangement and proto-matter geometric structure for primordial positroniums.

The mass ratio of an electron to a neutron is about 1 to 1,838*. The Positronium's rotational centrifugal force balances their internal electrical and gravitational attractions. *(Actual best known value for electron mass is one divided by 1,838.674704 compared to neutron and 1,836.142315 compared to a proton).

Proto-universe space and matter could, and may include other particles and anti-particles as well as positroniums, such as neutrinos and anti-neutrinos. If neutrinos, or other matter and antimatter, exist in the proto-universe, the structure would necessarily be much more complex. The deflagration process as described in this document demonstrates the feasibility of The New Universe Theory concept <u>within</u> the laws. It is certainly my desire that others, more abreast of current in-depth understanding of elementary particle physics should take this New Universe Theory concept processes, develop them as appropriate, and substantiate them with more detail analyses. However, primordial configurations that include other matter and anti-matter types are not expected to be found that could feasibly make up an alternative to this described proto-universe.

POSITRONIUMS IN HEXAHEDRON LATTICE CRYSTALINE STRUCTURE

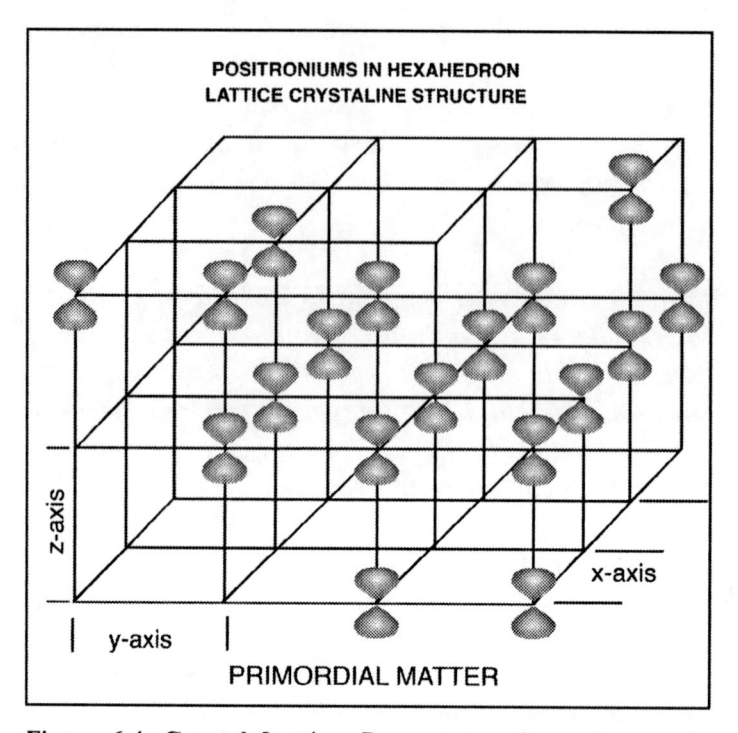

Figure 6.4. Crystal Lattice. Proto-matter is made up of positroniums in a monolithic hexahedron lattice arrangement. The 'Lattice' is hypothetical as in any crystalline substance, lattice is only an arrangement description Primordial matter structure has one positronium residing at every lattice vertex; (sketch shows only a typical few) they are individually stabilized by their rotation speed, and are stabilized in the hexahedron crystal by mutual electrostatic interaction as they continuously rotate in synchronization.

The proto-matter, being locked in balance in the lattice, is at an extremely low temperature, ...absolute zero. Temperature of a substance, by physical definition, is the energy level in that substance. In the physics of Kinetic Theory of Gasses, we learned the internal energy of a body of fluid is in reference to mean random momentums of the particles in the substance and that provides the energy (heat) to the sensing device for measurement. We know the positroniums cannot be moving randomly about or the proto-matter would be unstable; (i.e., exposed to contacts, and therefore annihilations). Fixed positions are required and are assumed for proto-matter. The proto-matter is stable in proto-space.

Rotational stability within each positronium is demonstrated as shown in the Figure 6.5 graph. A tendency, by any outside influence, to reduce the rotational speed of the Lepton pair in the positronium results in a shortening of the independent variable separating distance, (d), which in turn causes an increase in the rotational speed, (N), increasing the centrifugal force to maintain the separation. This increase in rotational rate occurs like it does for a figure skater. Because of the law that governs the conservation of angular momentum, figure skaters can increase their rotational speed by bringing their arms inward, adjacent to their sides; Then slow again when they extend their arms outward. All consistent with the laws of conservation of rotational momentum.

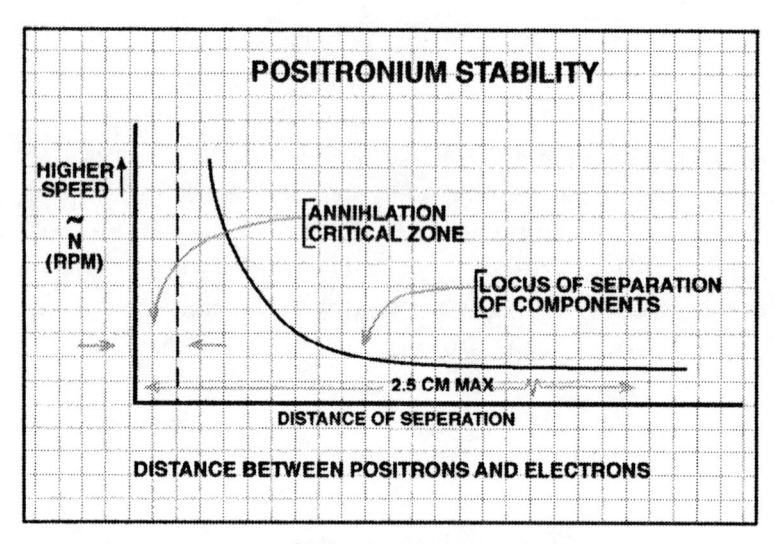

Figure 6.5. Positronium Stability. This "Stability Curve" illustrates the Physics principal for conservation of angular momentum within a body. The math equations for this relationship between rotating speed and separation distance are included in Appendix 2.0. The electrons and positrons are bound together by both gravity and electrostatic forces, and held apart by the balancing centrifugal force from rotation. The rotational speed of the positroniums are also stabilized by the influence of synchronization with rotations of adjacent positronium couples.

The intrusion into the proto-universe by the deflagration wave's electrostatic and gravitational forces cause positronium rotational perturbations, which become chain reaction, and the mechanism for the perpetuating annihilations of the deflagration front. Annihilations are the source of photons which combine to produce the elementary and sub-elementary building block mass particles that combine into cosmic radiation, nuclides, and chemical elements.

In chemistry, for describing geometrical fixed molecular separations in crystalline solids, hypothetical lattice structures are envisioned. The positronium 'molecules' *(positroniums are about 1/918th the mass of the smallest chemical element nuclide, the core of a hydrogen molecule)* reside at each and every vertex of the geometric lattice that defines the physical arrangement and uniformity of separation. These arrangement and orientation requirements are what chemists refer to as crystalline solid lattice structure of geometric solids. Common crystalline lattice configurations are tetrahedron, hexahedron, or octahedron.

Since spontaneous annihilations do not occur in proto-space, the rotating positron and electron of each positronium are of adequate, equal, and fixed spacings relative to the adjacent positron and electron couples during the continuing rotations. The rotations of these positronium dipole couples are locked in synchronization for continuous balance of attracting and repulsing electrostatic positive and negative ends of adjacent positroniums. In addition, they all must maintain uniform rotating speed to maintain the balancing centrifugal force between their paired elementary particles.

Lattices do not physically exist in chemicals or in positronium assemblages, these are simply references for envisioning geometric arrangements and separations of particles within crystalline solids. Solids are defined as substances of which molecules are not free to move about such as in liquids or gases. The molecules of physical solids can rotate and vibrate, however, for molecular solids, molecules are in contact, and that is how the transfer of heat is accomplished through the solid from one side of that solid to its other side. In proto-matter and proto-space, the positroniums are not in physical contact with next door neighbor particles. Their dipole axis of rotation must be the same throughout proto-space, so that dipole rotations can occur in synchronization in the adjacent perpendicular planes, but parallel to all other planes of rotation. Their electrical fields do interact, as these fields extend forever, providing the forces for stability of synchronization. The only lattice geometry that corresponds to proto-matter spatial requirements, and can maintain orientation requirements for stable positroniums are hexahedrons. The length of the axes legs in the lattice are defined by the density required to match the density of the universe. The value I have assumed for the universe's average density, considering all observed matter to this date is equivalent to eight protons per cubic meter. (For other estimates see Ref.(2)). In accordance with the first law of thermodynamics, the mass density on both sides of the deflagration wave must be equal, as the wave process can only synthesize and transform, it cannot create or destroy matter. Using our assumed average, the equivalent population of positroniums on the other side of the deflagration front is eight thousand (8,000) per cubic meter. The mass ratio between protons and positroniums is precisely 938.271998, but for convenience we use the approximate ratio of 1,000.

Also, the actual percentage of positroniums are currently unknown that are transformed into sub-elementary particles that are not integrated into neutron and proton production (such as mesons, muons, free quarks, neutrinos, and possibly other particles). Those particles become what some refer to as cosmic radiation, and prior to the NUT, their source has been unknown.

The equivalent universe density for positroniums of 8,000 per cubic meter results in lattice legs of 12.5 centimeters (4.88 inches). The axes of rotation are separated by 12.5 centimeters in two directions, and the positroniums on axes are separated by 12.5 centimeters. In this case the lattice cells are cubic hexahedrons. If in the future, astronomers surely will better define the universe average density, then lattice leg lengths of the New Universe Theory model must be uniformly adjusted to match the new values. The universe density will include the 'dark matter' (about 6 to 8 times more abundant than all directly observable matter). With inclusion of dark matter equivalent mass in primordial matter, the positronium separation is reduced by ½, which increases the cubic meter count by 8 times. Positronium separations are therefore 6.25 centimeters in all six directions.

It is interesting to note that the most stable and hardest crystalline materials known to mankind are also arranged in the six sided lattice shape of equilateral hexahedrons (usually cubes). Those substances include the hardest of compounds; Diamonds, Carborundum, and Tungsten Carbide. The chemical analogy is only symbolic.

All elementary particles from which all material things are composed have matter and anti-matter counterparts. Existence of anti-matter (positrons) outside the particle physicist's laboratories is further attested by the recent photo of *(Figure 6.6) "Huge clouds of antimatter (positrons) which have recently been observed, appearing above the core of the Milky Way galaxy." Apparently those positrons are being split from high velocity violently colliding matter, especially that orbiting and descending towards one or more black holes near the center of the Milky Way.*

(Read more interesting stuff about current knowledge on the center of the Milky Way Galaxy in Fulvio Melia's book on "the black hole at the center of our galaxy", see Ref. # 33)

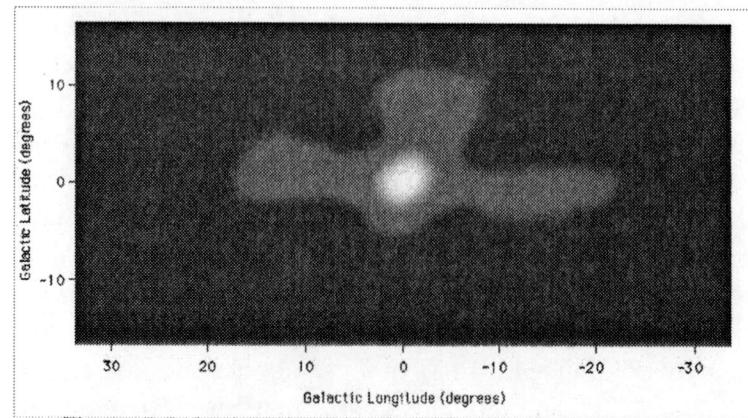

Figure 6.6. Antimatter Cloud. A ghostly cloud of positrons (antimatter) was recently discovered above and within the disk of the Milky Way Galaxy. Our solar system is also in the disk, but is about 18,000 light years beyond this cloud. We are about 28,000 light years from the center of this cloud and the center. This recently discovered cloud is glowing in gamma rays produced by annihilating antimatter particles; converting their masses to energy. Annihilation energy is emitted as gamma rays in photon pairs, each with energies of 511,000 electron volts. The 'OSSE' instrument on board NASA's Compton Gamma Ray Observatory has produced this map of the galactic center region. Annihilations show as gamma rays in the bright region in the vicinity of our galactic center at the central area and fainter horizontal emission from the galactic plane. Gamma rays are electromagnetic photons with very short wave lengths, shorter than x-rays. They are less than one angstrom in length, compared to visible light which has wave lengths from 3,100 to 7,700 angstroms in length. This large unexpected cloud of annihilation radiation extends about 4,000 light years into the disk, also extends nearly 3,500 light years above the galactic center. We are about 24,000 light years farther out, beyond the cloud. *Figure 6.6 credit to J. Kurfess at the Naval Research Laboratory, W. Purcell at Northwest University, and NASA.*

The characteristics for proto-matter defines a solid crystalline substance, with extremely low density by earthly standards. This, and the universe's average density are lower than the best achievable vacuum achievable in the most sophisticated laboratory on earth. Complete, 100% order and zero randomness is the condition for zero entropy. This is the environment and condition for the pre-universe of Stage I, before interaction by the reaction mechanisms of Stage II.

The extent of the reserve of the proto-universe and the amount of proto-mixture available appears to be unlimited, (maybe infinite; if there is such a thing as infinity, this may be it!). This primordial matter is an endless, three dimensional monolithic hexahedron lattice crystalline primordial structure of positroniums with continuing uniform density and particle makeup. This structural assemblage is described as **Stage I**. We do not presently know, and we may never know, or ever be able to intelligently estimate the overall extent of proto-space and proto-matter. Someone may someday develop a plausible theory for the source of primordial proto-matter.

This continuous Monolithic Positronium crystalline material occupies all of the volume outside of the universe. This could be stated as all the volume outside the NUT shell. The crystalline material has a hexahedron lattice-structure with a positronium located at each and every vertex. Until the deflagration wave (Stage II) intrudes, each positronium is rotating in synchronization with all others, this maintains the gravitational and electrostatic balance of forces for uniform separation and positronium stability.

Chapter 6-II ... Stage II Reduction Mechanisms

The universe which we can observe (or is available for observation) from earth or anywhere else in the universe is defined as <u>Stage III</u>. It could also be described as the matter inside and behind the NUT shell (Stage II). <u>Stage I</u> is the primordial matter of the proto-universe. <u>Stage III</u> is the result of the continuous flow of space and matter that is being processed through and by the <u>Stage II</u> deflagration front reduction mechanisms. This on-going process has been occurring for many billions of years. The <u>Stage II</u> deflagration front is made up of three consecutive concentric Sub-Stage processing shells of various thicknesses. The three zones of the reduction mechanism are phases that inter-lap and overlap, and are identified as <u>Sub–Stages</u> <u>IIA</u>, <u>IIB</u>, & <u>IIC</u>.

<u>Sub--Stage IIA</u> Annihilation wave: Positroniums become Photons.

<u>Sub--Stage IIB</u> Photons become Mass particles; Nuclides, and Isotopes.

<u>Sub--Stage IIC</u> Isotopes become Molecules, accreting into mist and Clouds of Dust and Gases. The Clouds

compound fragment, then condense, clump, and accrete into Stars, Clusters, Galaxies, some are larger galaxies containing Quasar cores.

*Each year these three **Stage II** "Reduction Mechanism" shells continue to move through the proto-space at light speed adding two (2) more light years to the diameter of the universe. Each and every second the shells add, in all directions, 186,000 miles which is almost 300,000 kilometers). The mechanisms continue for 3600 seconds per hour, 24 hours per day, 7 days per week all year long, and that calculates to be 5.88 trillion miles. This has been going on for more than 21.+ billion years. (The basis for this number is substantiated in Chapter 11). The universe is very big compared to we little humans, our little earth, and even our little galaxy. In the process there is a lot of matter (includes mass and energy) being added to the universe. The amount of space and stuff being added each year is **equal** to the amount of proto-space and proto-matter being consumed and processed by the deflagration front each year. The Laws of Physics require continuity and the laws further state matter can neither be created nor destroyed, thus requiring this equality. There are no other scientific options.*

Stage IIA

Annihilation wave: Positroniums (Leptons) become Photons.
Primordial Matter Transforms to Energy

When particles and their anti-particles (positrons and electrons are Leptons) are brought into contact, they annihilate each other and instantaneously all of their mass is transformed into energy. The energy radiates in the form of gamma ray photons which are electromagnetic waves. The deflagration wave is a spherical shell of annihilations, moving outward at the speed of the combining / fusion of positrons and electrons; it is made up of the annihilation generated electromagnetic radiation photons. For every meter of front propagation into the crystalline proto-matter, another 8,000 positroniums are consumed through each square meter of deflagration surface and these are annihilated into another 8,000,000,000 eV (electron Volts) of energy. This approximately corresponds to the amount of energy necessary to generate four neutrons and/or protons (the nuclei of hydrogen atoms). Fusion of electrons and anti-electrons (positrons) through the transmuting of positronium is the only conceivable source of fuel adequate to provide the huge amount of energy needed for powering the deflagration front and simultaneously producing the photons for feeding the mechanisms; And then converting into the mass particles which are the building blocks from which all of the universe's objects are made. (Annihilation of 918 positroniums (1,836 leptons) produce about one billion electron volts of energy, the quantity needed to generate one neutron). The front is believed to be propagating like a pulse jet with it's mean velocity infinitesimally, below the speed of light, as a time increment is needed for photons from the annihilations to accumulate into concentration quantities for precipitation into neutron and proton particles. *This possibly occurs intermediately into quarks and then into neutrons, and subsequently into protons (Hydrogen nuclides) and beta particles (electrons).*

The term "deflagration" was borrowed from a chemistry process as that type of combustion rapidly propagates through a combustible fuel and oxidizer mixture in a manner similar to the annihilation front, progressing through proto-space and proto-matter. Also borrowed from the chemistry field is the term "reduction mechanisms", the sequence of processes that ultimately leads to the formation of other products. These chemistry terms apply here as well.

An analogous acoustic shock wave moving through the earth's atmosphere is somewhat like the deflagration front. An acoustic shock wave momentarily, isothermally (no temperature change) compresses and decompresses the gas, and in the case of the acoustic shock wave, as it passes on through and leaves the atmosphere as it was before, except for some minuscule viscous losses. In the case of the Sub--Stage IIA deflagration, the wave's momentary impulses upset the tenuous balance between and within adjacent positroniums in proto-space, and sequentially cause particle to anti-particle contact; annihilations instantaneously occur.

The wave of annihilations continuously releases tremendous amounts of energy, and perpetuates the deflagration front at, and it is speculated that it is <u>not at</u>, <u>but is almost at the speed of light</u>, as more electromagnetic photons are produced. The photons, being electromagnetic waves, move at the speed of light, so they catch up and accumulate in the wave front. The spherical outer surface of the deflagration wave is the propagation of the combining / fusion of positrons and electrons. The magnetic and electrostatic impulses of these electromagnetic waves produce compression and reduction of space between and within the positroniums of the proto-universe. This upsets the balance, to initiate, propagate and perpetuate the chain reactions that continuously maintain energy production for fueling the <u>reduction mechanisms</u>.

NEW UNIVERSE THEORY WITH THE LAWS OF PHYSICS

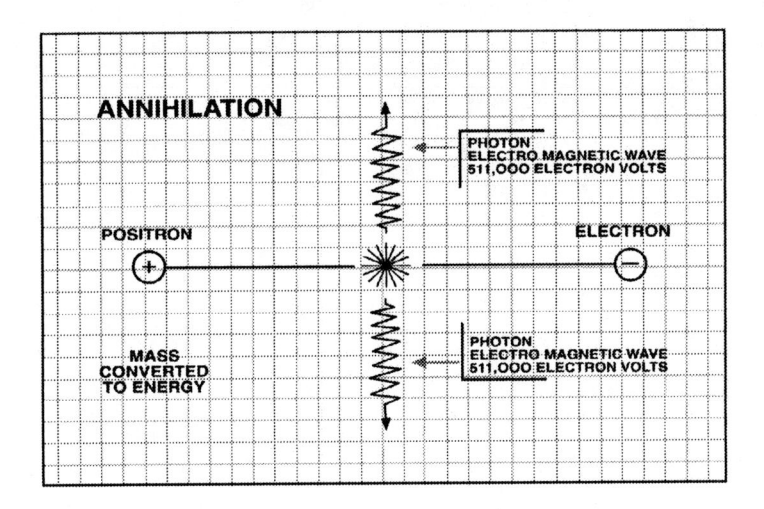

Figure 6.7 Annihilation. When a unit electrical charge comes in contact with another unit electrical charge of opposite polarity, the electrical charges combine and obliterate each other, but matter cannot be created or destroyed. Each of the mass objects are converted into energy in the form of one photon each. Photons are electromagnetic waves, and for positron and electron annihilations, the energy content of annihilated photons are each 511,000 electron Volts.

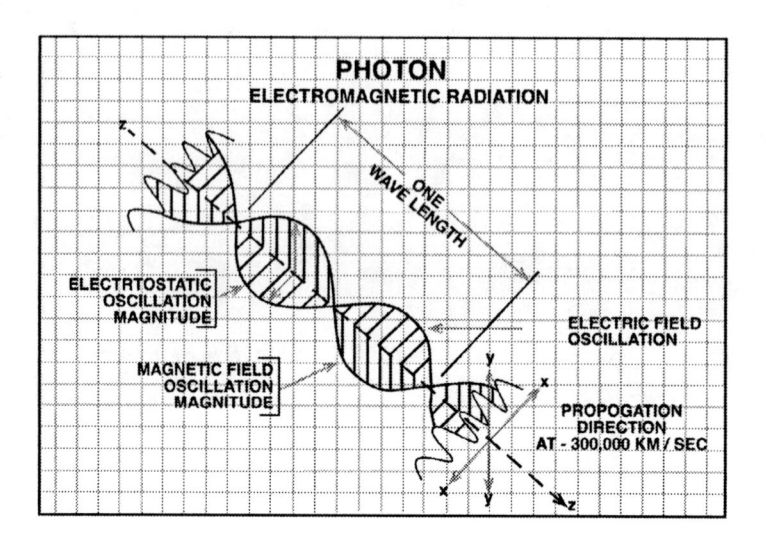

Figure 6.8. Photon. All electromagnetic radiation propagates in energy packets calle photons. (Sometimes described as corpuscles of energy) Electromagnetic waves all travel at the speed of light, as light is simply optically observable electromagnetic radiation. This energy travels in a straight line unless interacted by magnetic, electrical, or strong gravitational fields. As the energy wave travels it oscillates in two perpendicular planes, electrical in one plane, and magnetic in the other. Photons can have various energy content, however, positron and electron annihilations always convert into one photon each, having energy of 511,000 electron Volts.

Photon density builds and accumulates in the trailing part of the front from the annihilations, and where they accumulate into the energy quanta that combine and then precipitate into mass units. These transitions from energy to mass precipitants are possibly via smaller particles, e.g., quarks and more leptons, in the process of producing Hadron mass (neutrons, protons, muons, gluons, and other elementary particles.) Quarks are about 1/3 the size of Baryons. (Quarks are described in the Stage IIC discussion.) The precipitations are intermittent, of course, as the intervals for accumulations of the appropriate quantities of photons are also incremental.

The fact that photon (energy) converts to mass is of no doubt. Questions yet to be answered by particle physicists pertain only to the details of how?; and through what intermediate steps? We may not learn the answers to these questions for a long time, not until a lot of additional research and analyses are conducted. Particle physicists are conducting on-going research at laboratories like CERN and Fermi-lab, are resulting with theories about more particles known as Biggs, Bosons, Gluons, among others. We need not trouble ourselves at this time, as the <u>lack of detailed answers to these questions do not impede our ability to understand the New Universe Theory in general</u>.

The initial primordial mass density of the **Stage I** proto-universe does not change the reaction mechanism, it affects only the quantity of particle precipitations. We will eventually know the density of the proto-universe because the total matter into the front must be equal to the matter out; **the continuity <u>Laws of Physics</u> do not allow for any alternatives.** However, at this time, we do not have an accurate measurement of the density of the Stage III universe. The important reason for bringing up the point again is to emphasize that the density of the proto-matter and proto-space involved in the process, does not have a significant, if any, affect on the reaction mechanism processes.

We can speculate with logic to bracket the thickness of the **Stage IIA** shell. For each 12.5 centimeters of travel, every square meter of front intercepts sufficient quantities of positroniums (lepton pairs) to produce one hadron (proton or neutron). For each meter (100 centimeters) of deflagration front propagation, the photon accumulations are sufficient for eight (8) hadrons through each square meter area. Much study and analyses remains for analytical-particle physicists to determine and substantiate the concentration requirement. Therefore, for now, it is intuitively estimated that the **thickness of Stage IIA shell is more than 12.5 centimeters and less than 100 centimeters** (one meter).

Stage IIB
Photons become nuclide Mass particles: mostly neutrons.
Energy Transforms to Mass

Photons convert into neutrons, some of which quickly decay into protons and electrons (beta particles), plus some additional matter, all of which is precipitated into the universe's new space. This new space and mass is growing and traversing outwardly at near the speed of light. These products trail the annihilation front which also is traveling at almost the speed of light. (Almost, not at, because some of the Energy to Mass precipitants intermittently accrue and coalesce in increments, and mass cannot travel at the speed of light. (Also some accrue in diverse directions?).

Annihilations of **Stage IIA** result in primordial mass being converted to energy which occurs in accordance with the first Law of Thermodynamics, (conservation of matter) which states matter can neither be created nor destroyed. *However, all of the annihilation energy photon units are not likely to be converted back into mass, which leaves some electromagnetic* *waves to propagate through the universe and become what is currently referred to as background radiation.* The energies from the annihilation of 1,838 electrons or positrons (919 positroniums) are required to generate one proton. The rest mass of 1,838 electrons or positrons has a rest mass equal to the rest mass of one proton. (a proton is also the nucleus of one hydrogen atom, also referred to as one hydrogen nuclide). Based on some mass measurements by several astronomers, and with the laws of continuity, each cubic meter of primordial matter must include ~8,000 positroniums. Any energy in excess of that totaling the formed particle's mass equivalent, is transformed into other matter, i.e., neutrinos, leptons, muons, or simply remains as photon radiation to become additional background radiation. *(background radiation can be seen by simply tuning your antenna connected TV to a channel where no stations are broadcasting a signal. The snow on your TV screen is partially a result of the signal from background radiation.)*

Each photon is a unit quantum of energy that propagates at the speed of light; as a matter of fact, it is light, an electromagnetic wave oscillating electrostatically and electromagnetically as it proceeds at this highest possible speed; 300,000 Kilometers per second or 186,000 miles per second.

As previously stated, we do not know precisely how and through what steps matter is transformed from energy to mass, but we do know that it does so, and in accordance with Einstein's famous relationship of Energy equals Mass multiplied by the square of the velocity of light. (**E = M c²**). This relationship was discovered and defined by Albert Einstein in 1909, and is now accepted as one of the Laws of Physics, as it has been proven many times over. (At many laboratories, in high energy particle collisions, through analyses of nuclear fusion; also in nuclear fission reactor power plants around the world. It has also been proven through process analyses of radiation from stars.) Albert left us without telling us how this process works, only that it does.

The process requires annihilation of leptons and produces photons, and then through photon wave interference, coalescence and contractions, mass is reproduced, but this time as protons and neutrons. We can speculate on what appears to be a few intermediate steps of matter transformations involving intermediate size particles. Likely intermediate type of particles may be quarks. The energy-to-mass transformation would be more plausible and easier to comprehend, if the reduction mechanisms are actually through smaller steps rather than through the leaps and bounds. Accumulations of smaller quantities of photons seems more probable. Quarks, although never having been observed in the laboratory, are currently theorized by particle physicists to be the most elemental of elementary particles. Quarks are theorized to be building blocks for protons and neutrons, as well as for other Baryons.

A cursory-overview summary of current quark theory follows:

The currently thought most elemental constituents of mass are the Quarks. There are six (6) different 'flavors' and are labeled u, d, s, c, b, and t; (abbreviations for up, down, strange, charmed, bottom, and top. The names show a sense of humor within the physics community, as the terms have no relationship to the conventional meanings of the words). The d, s, and b, quarks have a negative charge that is 1/3 that of an electron; the u, c, and t, have a positive charge that is 2/3 that of a positron. Different flavored quarks have vastly different masses. Various quarks apparently combine into very specific arrangements which result in various other specific particles. Theory holds that protons and neutrons each contain specific combinations of electrons, positrons, some other small particles, plus three quarks, all of which add up to the masses and charges as known to be contained in protons and neutrons.

For brevity and simplicity of presenting the New Universe Theory concept, this discussion involves only conversion of (Positroniums) Electrons and Positrons as proto-matter, to Photon energy, and then photons converting to mass Protons and Neutrons. This process is valid, but further complexity of details are avoided, which are beyond and outside the objective of this work.

Subsequent to the generation of mass, the only process for slowing the particles' linear velocity is the conversion of linear momentum to angular and rotational momentums, which are induced by electrostatic and gravitational interaction between them. In accordance with Einstein's relativity equations, no mass can travel as fast as the speed of light, only near it. At these speeds, the interactive mixing and coalescence of these newly generated baryon particles occurs intermittently, and therefore this region / phase covers great distances, as velocity reductions and entropy grow slowly. However, early-on, before linear velocities begin to reduce significantly, the **velocity-enhanced-gravity** is adequate to fuse Hadrons into proton and neutron isotopes because paths are frequently diverted by adjacent particles' gravities. Some particles collide and fuse into nuclide isotopes, and some accrete into yet heavier isotopes. As they become larger the mixing rate and vortexing increases, linear velocity slows, and entropy grows to consume linear momentums. **Entropy** is discussed more comprehensively in subsequent chapters.

The thickness of the **Stage IIB** shell part of the deflagration front is of course, at this time, impossible to accurately calculate. This is the part of the deflagration front where the photons propagating from annihilations collect into specific quanta and then precipitate into mass particles, including proton and neutron particles. The first lone particles are all traveling at near the speed of light when they first form, and are separated by random distances dependent upon where the photons congregate and coalesce into various particles. The particles must migrate perpendicular to the direction of the speed of light motion of the front to coalesce with other particles. This coalescence is of course motivated by velocity-enhanced-gravity.

Logical analysis and description of the **Stage IIB** process is that the newly formed neutron and proton particles flow from the front like a mist in a sparsely populated fog. Actually, the front keeps moving outward at light speed, and the new particles lag and follow behind the front. "Out of the Mist" is the continuous fog bank of particles which are flowing at near the speed of light, and therefore their mass and gravities are <u>velocity-enhanced</u> in accordance with Einstein's relativity equations. The contiguousness of this cloud is short lived due to the individual particles' enhanced-gravities. Cloud fragmenting occurs at all levels, from individual particle pairs to groups of pairs, and groups of groups, up to the volume inclusions of several hundred million light years.

Transverse migration of particles, groups of particles, and groups of groups also comes from mutual attractions by the particles' electrical charges, but attractions are primarily by the velocity-enhanced mass and gravitation. The individual particle initial transverse travel distances, velocities, and travel times can be calculated, however. The individual particle initial transverse distances vary over the range of 5 centimeters to possibly 100 centimeters, and they must accelerate from zero transverse velocity to close the gap for accretion. The mutual attraction and coalescing forces comes from the high, near light speed, velocity corresponding relativity mass factors and the directly related gravitational force factors (discussed and calculated later). The complexity of the calculation of the distance over which this takes place makes it appropriate (in this document) for us to make an estimate in order to have a value for the shell thickness. My current estimate (based on analysis discussed later in this document) is that it will require a distance of travel somewhere between 10 and 10,000 times the initial transverse separation distances. The corresponding distance range for **the thickness of Stage IIB shell is between 50 centimeters and about 5000 meters, which is only five kilometers.**

Understanding is lacking about forces. This area of physics is in need of research and analysis for definition of the phenomena of forces. Some forces are strong, some are weak, some only attract, and some can attract or repel. There are none known that only repel. Some act over only short distances, and some act over distances as large as infinity. Most operate as a function of the inverse square of the separating distances, and yet some are simply indirectly proportional to distance.

The most significant force we are concerned with in understanding the New Universe Theory is the force of gravity. Gravity emanates from all mass, which is a variable as a function of the velocity of the mass. Albert Einstein defined how the mass, gravity, and velocity relates. The relationship is called the Einstein theory of relativity. However, the principle and phenomena have been proven many times and therefore should now be called Law, not theory. The relativity forces and masses are described in the next section which reveals how the elements are created by gravitational coalescence of protons and neutrons, not by high temperature (high particle velocities) and high momentum collisions. Velocity enhanced mass and gravity are quantified in the next section, in accordance with Einstein's relativity.

The following is presented only as an introduction to force phenomena; There are only four types of forces that physicists presently "presume" to exist in the universe. As you can see, this is an area of physics where much definitive work is yet to be accomplished. (Some proto-physicists could make a contribution to science in this field.). Forces that are today only vaguely described are:

Electromagnetic Forces: Includes electrostatic and magnetic.

Gravitational Force: 'At rest' Gravitational/inertial forces are the weakest of forces. Near the speed of light, Gravity is the largest force. (It is not clear if the mass or its gravity, or both, is/are the variable(s) that change(s) with velocity, but the result is the same)

The Strong Force: Nuclear binding. Considered the strongest, acting only over very short distances. Gravity can match this under certain conditions. (e.g., on neutron stars, and on quark stars)

Weak Force: This is closely related to electromagnetic force, such as beta (electron) ejections from nuclides; e.g., a beta particle is ejected from a neutron when it decays into a proton.

Stage IIC
Nuclides transform into molecules and coalesce into stars, clusters, galaxies
Particles Coalesce Into Observable Objects

Isotopes form from a continuous mist of neutrons. Some neutrons rapidly (within a few million years at speeds near light speed) decay into protons and beta particles (electrons), plus some other matter. (Near rest, like on earth, one half of all free neutrons decay in 10.25 minutes, but while traveling at high speed with relativity time dilation, the half life is much longer). The mist fragments into Clouds of all sizes, from a couple or a few neutrons, to gigantesque sizes.

With gravitational coalescence, individual nuclides within the clouds merge and fuse into isotopes; these huge clouds further fragment into smaller clouds, which continue to fragment into yet smaller clouds. Nuclides continue to gravitationally coalesce forming nuclides of several sizes of isotopes. Even some smaller clouds are still huge, they gravitationally collapse and contract and clump into stars; multiple clumps form star clusters, galaxies, and some coalesce into larger galaxies containing quasar cores.

Up front, where the deflagration front interacts with the positronium proto-matter, the annihilations are producing photons. Photons and the front are moving outward at near the speed of light, and therefore so are the mass objects when they form. But at slightly slower speed, as mass cannot travel at the speed of light, in accordance with Einstein's relativity equations (appendix). Subsequently, the only means they have for deceleration is through mutual attractions, collisions, and via numerous repeated near collisions, resulting in loss of <u>linear momentum</u> that transfers into mutual orbiting and <u>angular momentums</u>. Neutrons continue to fuse due to the high velocity enhanced gravity, but those that don't fuse into isotopes remain unstable and decay into protons plus an electron, (hydrogen nuclides), (a neutron's, at rest half life is 10.25 minutes). The fusing of neutrons stabilizes them, whether the fusion is with other neutrons, or into other newly formed nuclides of several of the lighter elements.

The NUT concept suggests that the multiple neutron nuclides without protons, are stable and make up about 80 to 90 % of all of the mass of the universe. These neutron nuclides are undetectable except for their gravity characteristics. They are non-magnetic, non-electrostatic, and are invisible. These neutron nuclides act like other mass objects but do not interact other than gravitationally as an inert gas around, and in the vicinity of galaxies and other interstellar objects. Some normally transparent hydrogen gas (not inert) is visible in the January 13, 2001 APOD picture titled "a sky full of hydrogen"; Visit the web site. It is my belief that, the stable multiple neutron nuclides are the dark matter astronomers have been searching for so fervently over the past several decades. Such huge clouds around galactic super-clusters provide the lense effect that makes light from more distant sources appear as if space were warped. This is the dark matter for which we have been looking; it is also transparent. But what is the difference?

The AAAS (American Association for Advancement of Science) recently (12 December, 2003) published in their Journal "Science", a paper on the subject of "The Hunt for Dark Matter in Galaxies". The author of the paper was Ken C. Freeman, of Mount Stromolo Observatory, Australia National University. His paper presents his analytical conclusion demonstrating the ratio of observable matter to Cold Dark Matter (CDM) is 1:6; CDM is 85.7% of the total. Ken Freeman's conclusion is presented with a graph that shows where the CDM distribution must be to correlate with other mass distribution and rotational velocities in and of the galaxies. Ken, the NUT offers you the answer, in a nutshell. Thanks for your timely paper.

Up forward with the front, infrequent collisions occur among particles trailing immediately behind the "Annihilation" and the "Transition of Matter". To get this process in perspective, the mass particles from **Stage IIA** and **Stage IIB**, are traveling at velocities similar to, but even higher than those produced by physicists using the most powerful multi-billion dollar linear accelerators and cyclotrons. These particles' are initially traveling almost parallel to each other and have low relative velocity collisions, so they don't fragment as in the case of man made particle colliders where the particles are traveling in diverse or even opposite directions. Instead, the side by side nearly parallel path particles, have velocity-enhanced gravity attractions, resulting in fusions because of gravity forces, (not momentum). Fusion is the combining of hadrons (protons and neutrons) into isotope nuclides. Fusions require either high temperature (~10,000,000 degrees K) and/or very high pressure. Temperature is only a measurement of the mean momentums of the particles in the subject fluid. The particle fusion forces in fusion reactors comes from the rate-of-change-of-momentum (**F = ma**) between two particles in collision. Velocity-enhanced gravitational forces (pressure between particles) bring particles together more peacefully and gently to accomplish the result of producing nuclide isotopes.

After a time when the linear velocities have reduced due to momentum transfers through collisions, near collisions, orbiting, and vortexing, the velocity-enhanced gravity weakens and becomes no longer strong enough to produce fusion. However, enhanced gravity is still strong enough to result in clumping into molecules and clouds. This continues throughout **Stage IIB, IIC** and into **Stage III**. The probability of collisions is small throughout Stage II and therefore only a small percentage of the particles generated are larger than Hydrogen nuclei (protons). The percentages of free nuclides as observed in interstellar space indicate that first generation stars begin their nuclear fusion with a ratio of about 27% helium (contains four hadrons) with most of the remainder detectable mass as hydrogen. A satellite named Copernicus carried a spectrometer and detected the interstellar space between us and star -Centaurus (only about 4.2 light years distance) contains deuterium at 40 parts per million (by particle), and this, by the Laws of Continuity, should be similar to concentrations from the **Stage II** generated / produced products. Nuclides other than protons and neutrons, were probably produced early on by the described collision and coalescence-fusion process. It is not expected that the annihilation photon concentrations could directly produce multiple component (hadron) nuclei. Also, it is <u>not</u> expected that even through collision/fusion processes that do occur, nuclides would result that are heavier

Bobby McGehee

than boron; Probably at most, carbon would occur, but rarely. The heavier and more complex the nuclide, the more rare its abundance. Heavier nuclides that do exist at this phase, would disappear from observation as they would become seed for clumping growth that culminates into stars.

Figure 6.10 is an excerpt from the "chart of the Nuclides" and is presented for those who want a cursory view about the various combinations of protons and neutrons in various nuclides. This abbreviated chart was cropped from the Chart of the Nuclides, which contains and defines over 3,100 currently (in year 2003) known nuclides. This chart was cropped to display only the lighter weight nuclides. The atomic number (corresponding to the number of protons), is indicated by the numbers on the left. The number of neutrons contained in the nuclides is indicated across the bottom of the chart. The basic elements are identified in the heavier bordered squares on the left of the chart. The boxes in the chart are the isotopes of the basic element nuclides. The atomic weights and the half-life of most of the isotopes are indicated in the boxes, but heavier isotope data are not shown in this chart excerpt. Those isotopes are produced in stellar furnaces and supernova explosions, and those processes are outside the discussion scope of this book.

Copies of the complete "Chart of the Nuclides" are available from Lockheed Martin at; <nuclides.chart@lmco.com>. The complete published chart includes <u>all</u> of the <u>known</u> nuclides of which some include as many as 107 protons, some with as many as 158 neutrons, and to date the heaviest known has an atomic mass of 258. Only a few of the lighter nuclides are formed in the deflagration reduction mechanisms, shortly (within about a billion years) after the phase where photon energy is converted to mass. The heavier nuclide are from fusions in stellar furnaces and supernova. The heaviest nuclide fusions that result in release of surplus energy is Iron, and therefore that is the heaviest element (nuclide) that a star's nuclear furnace can synthesize. Heavier than iron nuclides *are formed during supernovae (star explosions) when the explosive forces and velocities are so great that some of the nuclides are jammed together into larger-heavier elements. When this happens, they absorb some of the explosion energy that produces these heavy, often unstable nuclides. Since these elements exist on our earth, we know our solar system evolved from a 2nd or 3rd generation star. We have learned that we can build nuclear power plants by concentrating certain of these heavier unstable elements and recover some of the energy they absorbed from their parent supernova.*

To recap, it is believed that the only nuclides that are generated during the **Stage II** *reduction mechanisms are those up to the atomic number of ~5; Boron, which contains 5 protons and 5 neutrons for a nominal atomic mass of 10. Someday, a computer model will allow these estimates to be substantiated or adjusted.*

In all of Stage II, the principle contributor to the coalescence and collisions that generates the nuclides are the velocity-enhanced gravitational forces. Mass is first formed with very high values compared to the rest mass as defined by Einstein's relativity equations. The results of high mass are accompanied by corresponding high gravity. *These particles at near the speed of light have higher mutual gravitational attraction by many factors (multiplier of their rest mass gravity force). It is of paramount essence that these force factors be properly included in the preparation of computer simulation models of the New Universe Theory concept. (the velocity factors from Einstein's equations have been proven many times at every particle linear accelerator and cyclotron and linear accelerator in the world). The actual magnitude of calculated velocity enhanced mass factors continue to increase without limit as the objects' velocity values are closer to the speed of light. The following table presents gravity values for high velocity, but stops where it does because gravities higher than those identified with a double asterisk (**) correspond to* *a quark star's surface, which is an incipient black hole. The only reason the real numbers don't go to infinity is because the initial velocities of the photon-generated particles are infinitesimally below the speed of light. Size of the increment below the speed of light is not known, only that it is small. (equation and further implications are discussed further in Appendix II).*

The following paragraph is a brief speculation about possible future research studies: A computer model of the NUT will require some assumptions about the initial velocity of the precipitated particles so that velocity enhanced gravity between particles can be quantified. Tangible substantiation, and maybe verification, of the New Universe Theory is almost at hand. It is interesting and fun to speculate about information that is potentially available through some yet to be conducted innovative research. Indirect measurement of deflagration front velocities, and maybe the initial mass precipitations, may be possible one day in the near future. If part of the microwave background data, as published in 2003, from the WMAP satellite is from excess positronium annihilation gamma rays that did not get included in the initial mass precipitations, measurement of their red shift might give us an indirect look at the speed of the deflagration front. Positron rest mass annihilation spectral distribution reference data is available from other sources. The objective of the WMAP survey was to look at background microwave energy distribution and therefore data was not acquired in discrete frequency spectral distribution which is required for deflagration red shift analysis. WMAP data was acquired in frequency bands, (e.g., wave lengths from one cm to five cm), which does not lend itself to red shift spectrographic analyses. When appropriate studies are conducted for acquiring spectral data of the microwave background, red-shift analyses could reveal extremely important and interesting velocity information, to quantify velocity enhanced mass and gravity. Of course much other information such as the propagation and increasing rates of vorticity and corresponding entropy are also needed for comprehensive modeling. For now we have to rely on theory alone, and all speculation of processes are credible as long as we stay strictly within the Laws of Physics in our thinking.

The velocity enhanced gravity calculations are discussed more in the appendix, but comprehending the enormity of the gravity forces can be recognized by perusing the Table 6.9. By using the following table, it is not necessary for readers to understand the relativity equations or accomplish any of the implied calculations and analyses to evaluate the magnitude of these velocity enhanced gravitational forces. High compressive force between particles is essential for fusion of neutrons and protons. Pursuing this vein of thought and analysis is exciting to cosmologists and other physicists. Research and analyses are needed and the depth of one's pursuit is open and up to the individual reader and researcher..

Multiple neutron and proton fusions are the sources of all of the nuclides, except for simple Hydrogen (single mass unit proton). Most other elements come into existence through this enhanced gravity process and subsequently through further fusions into heavier nuclides (elements) in the high pressure furnaces of stars and their subsequent supernovae. Only a few, if any, are fused from the momentum force in particle collisions, as collisions usually result in fragmentation (fission) of the nuclides rather than fusions.

Velocity Enhanced Gravity		
v / c Velocity divided by light speed (%)	Mass Multiplier Factor (F_M)	Force multiplier Factor, (Weight Factor) (F_g)
0.0	1.00	1.0
25.0	1.03	1.07
50.0	1.15	1.33
75.0	1.51	2.28
80.0	1.67	2.78
90.0	2.29	5.20
99.0	7.09	5×10^1
99.9	22.4	5×10^2
99.99	70.7	5×10^3 (5 thousand)
99.999	223.6	5×10^4
99.9999	707.1	5×10^5
99.99999	2,236.1	5×10^6 (5 million)
99.999999	7,071.1	5×10^7
99.9999999	22,360.7	5×10^8
99.99999999	70,711.2	5×10^9 *
99.999999999	223,607.3	5×10^{10} (50 billion)
99.9999999999	707,112.1	5×10^{11} **
99.99999999999	2,230,607.6	5×10^{12} (5 trillion)
99.999999999999	7,071,120.1	5×10^{13} (500 trillion)

Figure Table 6.9. Velocity Enhanced Gravity. This Table illustrates the high gravitational attraction forces between the particles generated at very high velocities. Particles are formed from coalescing photons which originated from annihilating positroniums. The factors are calculated using Einstein's Relativity equations which are presented in the Appendix. Gravitational attraction between all objects is very high while the objects are traveling at speeds near the speed of light. Forces are more than sufficient to coalesce and 'fuse' the particles from individual neutrons and protons into nuclides which incorporate from two, to a dozen or more such hadrons. Enhancement factors are higher than this table illustrates at higher velocities; gravities higher than those identified with a double asterisk (**) correspond to the surface of a quark star, which is an incipient black hole.

NEW UNIVERSE THEORY WITH THE LAWS OF PHYSICS

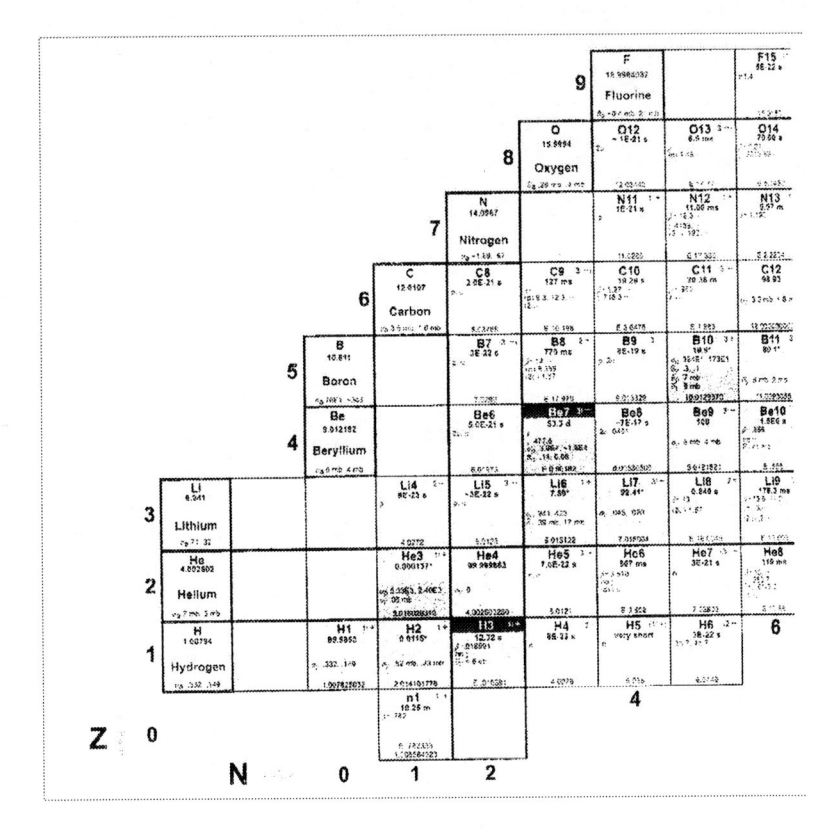

Figure 6.10. Nuclides. This cropped chart of nuclides displays only the lower mass nuclides; this is believed to identify most if not all of the nuclides formed by the NUT deflagration wave. Nuclides are the cores of atoms without any orbiting electrons. The atomic number (quantity of protons) is shown on the left; and the number of neutrons in these isotope nuclei are listed across the bottom. Atomic number defines element. The number of neutrons in each nuclide establishes isotopes of that element. All elements have at least a few isotopes. Each nuclide is shown in one of the boxes in the chart. Larger atomic number and heavier nuclides are formed either in the nuclear furnaces of stars, or in the explosive blast of supernovae when the heavier nuclides are dispersed into space. (A complete chart with all of the more than 3,300 known nuclides, as published by the General Electric company, can be purchased through <www.ChartOfTheNuclides.com>).

It is theorized as part of the NUT the multiple neutron nuclides are stable, which means they have an 'at rest' half life of at least 20 billion years. According to an article in the December 2003 AAAS "Science", the invisible halo around the Milky Way Galaxy contains about 20 times as much mass as all visible mass. The NUT implies probable answers for the missing mass mystery. Multiple neutron nuclides (Mnn).

Fusions require very high compression forces and the table on enhanced gravity reveals most fusions occur before velocities have slowed to speeds still faster than 99.99% of light speed. Accretions through enhanced-gravity coalescence produces atoms heavier than Hydrogen and Free-Neutron nuclides (both single and multiple). The gravity is still very enhanced below that speed of 99.99% required for fusions, and it continues to be instrumental in producing more orbiting and the transfer of linear momentums to rotational momentums. The much more numerous 'near collisions' divert the newly formed, high velocity nuclides from their original outward path. Linear momentums are thereby converted to angular momentum and rotational movement. In time, all particles, are linearly decelerated through transfer of momentums to angular momentum. As the result of their mutual attractions by both electrostatic and velocity-enhanced-gravitation, they are often caught in mutual orbits, then compound vortexes form and interact.

What causes vortexing? How and why do vortices form? When two or more particles are attracted together, their antecedent velocity component magnitudes and directions are never exactly equal nor perfectly aligned; consequently they collide or gravitationally interact with spinning, orbiting, or changes of their directional paths. When quantities of objects or groups of mutually orbiting cells come together, the resultant revolving is called vortexing.

***Alan Shapiro**, considered by many as the **father of modern thermodynamics**, once said "Old vortexes never die, they just fade away". (This is the saying that WWII General Douglas MacArthur borrowed and changed slightly for his retirement speech). In the higher density of fluid mechanics and kinetic gas theory, Shapiro's fade-away-phrase fits, but in the rarefied environment of space, the individual vortexes continue to combine and grow, apparently forever.* **(Never fading away)**

Matter enters into clumping after precipitating from the combining of photons into the mass forms (both baryons and leptons) in the preceding **Stage II B**, (photon-to-mass "mechanisms"). This **Stage II C,** starts at a velocity somewhat lower, many millions of light years distance behind the 'front', but continue following the front at near the speed of light. Coalescing and clumping collisions at these high speeds are the producers of isotopes and nuclei of the light elements (includes all hydrogen isotopes such as deuterium, tritium, as well as atoms of helium, lithium, some beryllium, maybe a little boron, and at least a few heavier nuclides such as carbon isotopes).

This accretion process continues until the primary decelerations at the beginning of **Stage III** are caused by secondary, and tertiary vortexes and by coalescence of vortexes, where relative velocities of the particles within a given vortex are low and gravitational attractions act more effective. Electrostatic attractions contribute to some of the molecular clumping but velocity enhanced gravitational attractions are primarily responsible for fusion-subsequent vortex formations. An important factor here is that as the particles enter a vortex, their aggregate linear velocities decrease. Then the relative velocities between the particles within a vortex become low, and the clumping forms molecules and groups of molecules, including clouds of dust particles. Large vortexes eventually form and coalesce with all sizes of mass particles ranging from protons, nuclides, isotopes, molecules, dust, and globs of mixed particles. Many vortexes are large enough to be hosts for several levels of smaller vortices, long before they lose significant gravity enhancing linear velocity. It takes several million years for velocities to decrease to .1% below light speed, and even then gravity will still be 5000 times rest gravity. (Illustrated in Figure 6.9). Vortices coalesce into larger vortices, many times over until the vortices are huge and are destined to become proto-galaxies. Interactions continue, especially as the relative velocities of the particles and smaller vortices are more easily coalesced to form planetesimals and brown dwarfs. Brown dwarfs continue to collect more mass and eventually become burning first generation stars of various types, depending upon the quantity of mass they continue to collect as they pass near and through clouds of dust and gas. Very large multiple neutron nuclides will never be found or observed, because they are the initial seed and core attraction for star size accretions.

Molecules start forming in Stage IIC and continue forming into Stage III where the velocity enhanced gravities are insufficient to produce fusions, but combined with electrostatic 'valence' attractions, are quite adequate to produce molecular combinations of atoms and isotopes. There have been several hundred chemical molecules detected in interstellar space (maybe thousands by now). Some of these have been reported in Astronomy texts such as Snow's "Exploring the Dynamic Universe" who in 1988 reported 66, some very simple such as the two atom molecule H_2, and others that are quite complex with as many as 13 atoms per molecule.

In early times of Stage IIC some large gas and dust clouds form with only small size vortexes, and with time and distance, these clouds coalesce into more compact groups, they form stars of various types and without the larger relative angular velocity compared to their neighbors as in the proto-galaxies, clouds rapidly coalesce into stars. These stars are the first to collect enough material to spontaneously auto-ignite and form the first star descendants trailing behind and from the deflagration front. According to conventional star age-dating studies, these stars ignite several billion years before those within the larger proto-galaxies vortex which take about another 3 to 5 billion years till their birth. They also eventually gather enough mass to enter the society of light generating stars, and collect into groups called globular clusters.

The thickness of the Stage IIC shell of the deflagration front is difficult to pin down, as are the thicknesses of the other Sub-Stages. Stage IIC begins when the first protons and neutrons combine, and are still flowing mostly almost parallel and outward, following the front, still at about +97% of the speed of light. Nuclide flow starts near the beginning of this Sub-Stage and during the transition through this shell, the nuclides coalesce into accretions of brown star mass clumps. Near the end of IIC, the flow turbulence and vorticity grow rapidly which provides for many collisions and accretions as well as mutual orbiting. The exit flow from Stage IIC is where the first stars are accumulating enough mass to begin to self ignite and start their nuclear furnaces. This is the region where the farthest away observed combinations of stars (Quasars) have been seen, which by our definition, is the beginning of Stage III.

It is anticipated that over the next several years, there will be much speculation and subsequent theories with computer/math models by astrophysicists as to how the front trailing mass accretes and star formations grow, in, through, and from the deflagration front.

Synopsis of Sub-Stages II A, II B, plus IIC

Synopsis is in present tense,... the universe is growing 'as we speak'.

The total travel distance of Stage II A + B + C corresponds to the distance traveled by the universes' objects in decelerating from light speed down to about 96% or 95% of light speed. With estimates of the deceleration rates in hand, it becomes possible to estimate the Stage II shell thickness.

During matter flow through Stage II, the proto-universe is evolving from primordial matter stored in positroniums and is transforming into the matter that is the observable objects and energy of Stage III. The transition/evolution process starts with the annihilation of matter and anti-matter, and promptly proceeds from annihilation produced photons into precipitation of mass particles. From the early existence of mass particles in the universe, their speed is very high, they are traveling mostly parallel and outwards, away from our older universe. Decelerations of particles are slow as it is almost totally dependant on vortexing and mixing. Mixing results in entropy increases throughout all of Stage II, (*also continuing throughout all of Stage III*), which occurs over a very long time period and distance, (billions of years in time, and billions of light years in distance). In addition, occasional collisions and accretions contribute to the conversion of mechanical energy (linear momentum) into both heat and the growth of entropy. All heat from collisions (mechanical energy converting to thermal energy) accumulates in the clumps and eventually, along with gravitational compression, contributes to star ignitions. Early on, in the evolution of matter through Stage II, a <u>mist</u> of particles which is a continuous cloud propagating forward while following behind the faster moving annihilation front. Due to high mutual gravitational forces between all particles in the cloud, the cloud rapidly contracts and fragments in a compound manner into regions that later contract and coalesce into stars, while the larger scale groupings of compound fragments collapse into galaxies. Almost all of the collapsing and clumping occurs with vortexing.

However, there are only a very few cloud fragments gravitationally contract without vortexing, both within the smaller clouds and within their 'parent' intermediate size cloud fragments. In these intermediate size units, without vortexing and centrifugal forces, the stars form sooner, and the stars oscillate through the geometric center of gravity of their group (rather than revolving around their combined center of gravity); these intermediate size non-vortexing star groups form what have become known as globular clusters. Globular clusters are usually captured later by the larger cloud-fragment descendant galaxies which evolve more slowly through larger scale vortexing. (Ref 39, "Globular Cluster Systems" by Keith Ashman and Steven Zepf; 1998.)

During this early development of the New Universe matter, much time is required for the collisions and vortexing plus clumping to achieve enough velocity reduction and simultaneous accretions to achieve the size needed for star ignitions. This estimate, based on a graphical illustration, is over the period of time is more than two billion years of time, and over a concurrent but longer distance, about two-and-a-half-billion light years.

In subsequent chapters of this document graphical illustrations are presented which are the source of both time and distance estimates. We know most of Stage II, (IIA + IIB + IIC), takes place during deceleration of mass from the speed of light down to the lower speed range of 96+ to 94+% of the speed of light, which is the farthest and fastest at which any object has ever been observed; (At a red shift of 27, which calculates to 96+% of the speed of light). Due to overlap between Stages, and by definition, Stage III starts at this or at a slightly higher speed. *(Five percent (5%) of light speed corresponds to an increment of almost 15,000 kilometers per second.)* The time and distance over which this Stage II occurs can presently only be graphically estimated.

When some extensive computer modeling is accomplished, from some yet to be acquired spectrographic survey maps of 'background' radiation, and then Doppler shift correlated with annihilation gamma ray wave lengths, we can then begin to get a clearer picture of the perimeter of the universe. Red-shift spectrographic maps, along with subsequent analyses will reveal many more new astounding facts. (*A recent significant achievement is the 2003 published WMAP sky survey which concluded in a map showing the variance of energy; A revealing result. What we now need is a map of the variance of microwave background in more discrete narrow band wave lengths*)

In conclusion of this Chapter: The synthesization of matter starts with the annihilation of lepton masses which converts to large quantities of photon energy. The processes and all of the steps through which this occurs is not yet fully understood. Elementary particle physicists are continuing to conduct research and gain knowledge, but most of their experiments are investigating the conversion of mass to smaller matter. We need more research data in the fusion of matter, most importantly photon to mass studies. When energy precipitates into neutrons with high, near the speed of light velocities with enhanced gravity, some neutrons (apparently about 85.7%) fuse into neutron nuclides before they individually decay into protons and beta particles (electrons). Many of these protons and neutrons, through their velocity-enhanced gravity, coalesce and fuse into a variety of multiple hadron nuclides (before deceleration to slower

than 99.99% of light speed), which apparently stabilizes them and decay ceases. The multiple hadron nuclides are more rare for the heavier nuclides, but some appear up to 10 or 12 mass units (Boron and Carbon). It is possible that some even heavier than carbon may be produced.

Stage III is the subject of the next Chapter, and it is there in the universe where most larger elements and isotopes come into existence, being processed in the nuclear furnaces in cores of stars. Nuclear physicists have proven that fusions producing heavier than iron nuclides absorb more energy than they release, and therefore fusion is no longer possible in the stars' nuclear furnaces. The heavier nuclides are produced during high pressure and high temperatures that occur inside and during supernova explosions. The only ones we get a chance to observe are of course only those that are expelled or formed in the out-flow part of the supernova. Further particle evolutions beyond supernovas are from fission of some of the larger unstable particles which convert some of their mass back into energy as they decay into smaller nuclides. Man made nuclear reactors control the rate of fission and recover some of the "stored up" earlier supernovae energy for purposes to serve mankind.

Note: When reviewing the galley for this book, I made myself a note at the start of this synopsis; "Boil it Down!" however, each time I read through it, I find the story so exciting I can't bring myself to delete any of the message.

Chapter 6-III ... Stage III
Central Universe
Potentially 'Observable' Matter; Objects and Energy

'Observable' refers to matter in existence for present or future observation, through use of optics or other electromagnetic radiation, gravity, neutrino, or other yet to be developed technology with which in the future we discover and develop with which to "see".

Stage III is defined as the observable universe. Anyone who knows enough about the heavens to call themselves an astronomer, including most amateur astronomers, are knowledgeable enough that they can explain star names, and their locations in the sky either by direction or by constellation. Most of these people can credibly describe "what is out there" in the observable universe.

My purpose in discussing the **Stage III** observable universe in this document is to relate the **New Universe Theory** concept to the existing universe and reveal how the universe and all potentially observable objects actually came into existence, as well as to why our universe evolved as it has. In so doing, this Chapter 6-III includes a medley of brief discussions about objects and phenomena compared to present day explanations and understanding, which was based on then accepted (BB) explanations for the origin of matter. To do so requires reflecting back into Stage IIC reduction mechanisms as they overlap into Stage III.

Eloquent descriptions and vivid photographs of the observable universe are available in many astronomy books, student texts, technical papers, and periodical publications. Recommended examples are: A coffee table book put together by Timothy Ferris titled "Galaxies"; Fulvio Melia has described current (2003) knowledge and understanding about the core of our galaxy in his book "Black Hole at the Center of Our Galaxy"; A comprehensive text, with many color photographs of observable universe objects are provided by Jay Pasachoff's field guide "Stars and Planets"; A comprehensive Astronomy text is, "Exploring the Dynamic Universe" by Theodore P Snow; Wil Tirions' "Sky Atlas" provides a map and master guide for finding observable objects, a must for all levels of astronomy viewing, amateur and professional; A couple of good monthly magazines that specialize in astronomy and cosmology are "Sky & Telescope" and "Astronomy". While reading, one must keep in mind these cosmological literary works were written when the Big Bang was still accepted, and therefore most are obsolete in some respects.

Reduction mechanisms and generation sequences of the New Universe Theory are the phenomena that are generating the overall observable universe. This NUT explains how observed objects and phenomena fit into the universe consistent with the Laws of Physics. First generation star ignitions begin within about 2+ billion light years behind the deflagration front, after adequate quantities of material accumulates and coalesces, as described in the preceding chapters. The collapsing of nuclide cloud fragments provide focused gravitational attractions that collect and concentrate adequate quantities of mass to produce internal pressure and heat to ignite thermonuclear fusion furnaces, and stars are born. The temperature required to start fusion is approximately 10,000,000 degrees Kelvin (Celsius + 273). Fusion force comes from nuclide contacts with sufficient force for nuclides to penetrate each others' "surface tension". The initial star ignitions result from velocity enhanced gravitational forces as well as from collision momentums converting to force. Subsequent star ignitions, at greater distances from deflagration front (and therefore at lower velocities) require a larger quantity of nuclides to provide the gravitational pressure and temperature for fusion ignitions. Later, second and third generation stars have less velocity enhanced gravities and therefore are more dependent upon larger quantities of nuclides for ignition. This is the type of star birth we see in nebulae, which consist largely of supernovae remnants.

This Stage III chapter presents a 'first hand' observable example of on-going universe development. Near-by Orion Nebula illustrates and typically shows universe development mechanisms are forever on-going processes. This convenient observation within our own little Milky Way galaxy, reveals second and third generation stars being born out of nebulae, large interstellar clouds of gas and dust, which is mostly remnant debris from earlier supernovae. Within the Orion Constellation the Orion Nebula is one of the most spectacular; it can be observed with a good pair of binoculars, as it is only about 1,500 light years away. **Direct observations are exciting and personally rewarding, that is why so many people are taking-up the hobby and becoming amateur astronomers. Direct-first hand seeing helps one to realize these mind-boggling Astronomy phenomena are real and actually exist. The photographs in this chapter are to illustrate what you and everyone can see for yourselves, first-hand.**

Color photographs are avoided in this book to allow this work to be produced more affordable to achieve wider spread distribution to many more inquiring minds. For colorful, beautiful and enlightening views of astronomical objects, many references are available. NASA has provided a valuable service by providing the internet site "Astronomy Picture Of the Day (APOD)". Another photograph is added to their site every day, as has been done since June 20, 1995. All are available for viewing and study, and copies can be printed for personal perusing. Frequent visitation of this site is recommended. All of the astronomy photographs shown in this document were down-loaded from the internet, and are printed (with permissions) but only in black and white. Publicly available astronomy views are available free for personal use to anyone with an internet connection. If you do not have direct internet access, APOD is available to anyone through most public libraries. Type into the search window the code name below, then explore and enjoy! <http://antwrp.gsfc.nasa.gov/apod/archivepix.html>.

Fascinating and astounding seem to be inadequate words to describe the magnitude and breath taking processes that are going on in our, as well as all other galaxies. The people at NASA have provided and shared with us some of the things that have been observed by Hubble Space Telescope (HST). Access to the internet and observing a video clip of phenomena gives us a beautiful experience, and it is almost incomprehensible. Downloading this video clip takes a few minutes after you go to the site, but it is well worth the wait. Type into the search window: <http://wires.news.com.au/special/mm/030811-hubble.htm> .

The photographs are accompanied by atmosphere music, so be sure your speakers are turned on. As stated by the friend that sent this to me said "the music is beautiful and the pictures are awesome". Stars that exploded in supernova to produce the nebulae of gas and dust that makes up these fascinating clouds are very large; some as much as 500,000,000 times the mass of our sun. These dust cloud nebulae that are photographed by HST are in our own Milky Way Galaxy. Similar activities are in process in all galaxies.

The size of the giant red stars as described in the foregoing paragraph is said to be 500 million times larger than the mass of our sun. Really big, not just a typo. After viewing the nebulae shown in the video from the HST, and recognizing that all the mass in each of these nebula of gas, dust, and debris came from an explosion supernova from only one, (some from a few) such stars, the size is acknowledgeable. The star that is considered the shoulder point in the Orion constellation is one of these supergiant red stars. This star "Betelgeuse" (Figure 6.12a), large at it is, could only be viewed from earth as a single point with the most powerful telescope until about 1995, when the Keck telescope first became operational. When one of these behemoth stars does explode in supernova, over half of its mass remains in place as the core of another star. The huge supernovae viewed by HST are all within a small sector of our Milky Way Galaxy. Also shown in that video are several far away galaxies, all of which are populated with stars, nebulae, and therein stars in production. With all this matter strewn about in various configurations, the universe is so large, that when considered on an average basis, the Stage III universe still has only average density of about eight (8) protons (hydrogen nuclei) in a cubic meter. Including dark matter,(compound neutrons) the density is less than eighty (80) per cubic meter. This is a better vacuum than mankind can produce here on earth.

Back to things we can observe first hand in the night time sky with our own eyes. Many objects are observable with the unaided eye, and many more are observable with binoculars. Viewing through amateur telescopes is even more awe inspiring.

Bobby McGehee

Figure 6.12a. Orion Constellation. Probably the second most readily recognized of the 88 Constellations in planet Earth's night sky is Orion, the hunter,. Cool red giant Betelgeuse is the brightest star at the upper left. Otherwise Orion's hot blue stars are numerous, with super giant Rigel balancing Betelgeuse at the lower right, Bellatrix at the upper right, and Saiph at the lower left. Lined up in Orion's belt (left to right) are Alnitak, Alnilam, and Mintaka all about 1,500 light-years away, born of Orion's interstellar clouds. And if the middle star of Orion's sword looks reddish and fuzzy to you, it should. It's the stellar nursery known as the Great Nebula of Orion. *(This Copyright photo appeared in the 2003 February 7 APOD. Reproduced with permission from astronomer Matthew Spinelli).*

Below Orion's belt, the second star in his sword is the Orion Nebula, a factory, not just a fuzzy single star, as it appears to naked eye observers. To quote a professional astronomer, "Few astronomical sights excite the curiosity about stellar birth as does this nearby stellar nursery known as the Orion Nebula." This nebula's glowing gas surrounds hot young stars being born from this immense interstellar molecular cloud. This nebula is a large cloud of gases, most of which are the result of prior supernovas. The numerous stars that are in production within the nebula will be third generation stars, like our sun. Other stars are in formation process as they are gathering material to start their own ignitions, and also, more mass is adding to the ones already in operation. The dozens of stars forming in the Orion nebula are grouped together in what is, or will be known as, an open star cluster. Open clusters revolve within our galaxy at various distances from the core of our galaxy, like the Pleiades. In early development of every region of Stage III, most nuclide clouds are much larger (many millions of times larger) than what we observe in the Orion nebula.

Figure 6.12b. The Orion Nebula. A cloud of dust and gases, most of which are the products of one or more former huge supernova. This object provides us a classroom, giving us a view of star formation progression. A supernova is the explosion of a star that has burned much of its fuel and thereby has lost much of its mass; when such a star loses sufficient mass to reduce its gravitationally imposed contraction pressure, the outward pressure produced by the star's internal nuclear reactions' rapidly overcomes the gravitational pressure, and the supernovae explosion ensues. The larger the original star, the faster it burns, and has much material to expel when it goes into a supernova. The resultant gas, dust, and debris cloud produces future generation descendant smaller stars. In addition, there are also significant amounts of virgin hydrogen in the interstellar medium plus the unburned hydrogen in the supernova explosion debris for them to accrete, and start many other 'star life cycles'. *(Credits for the photo; 2MASS Collaboration, U. Mass., and IPAC. Mosaic by E. Kopan)*

New star formation is more evident and apparent in the next figure when this closer view of matter deep within the Orion Nebula is scrutinized. The four brighter stars to the left of center are often referred to by astronomers as the Trapezium. (a geometric term for a figure with four unequal and unparallel sides)

Bobby McGehee

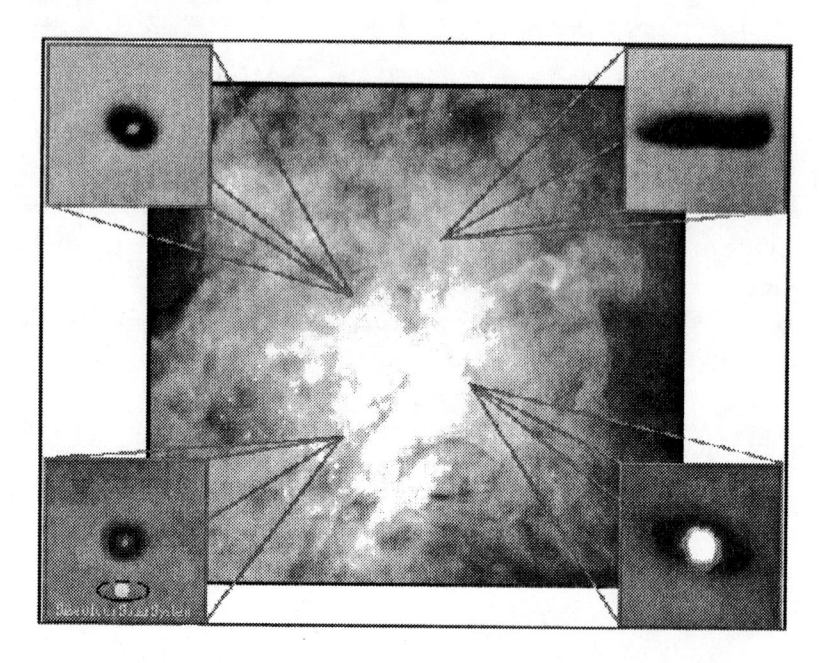

Figure 6.12c. Planetary Systems Now Forming in Orion. One of the more interesting of all astronomical nebulae known, is the Great Orion Nebula. Inserts to the mosaic show several planetary systems in formation. The bottom left insert shows the relative size of our solar system. The Orion Nebula is located in the same spiral arm of our milky Way Galaxy, as is our solar system. It is amazing to realize that the supernova of earlier stars provided such huge rich gaseous mixtures for incubation and birth of so many descendent stars. *(This photograph is with permission; credit to NASA and C. R. O'Dell, Vanderbilt U, and S. K. Wong. Rice U, WFPC2, HST. This figure, was first published by APOD December 7, 1996)*

The Trapezium stars in the Orion Nebula, at a distance of 1,200 light years, are in a pattern remarkably similar to those found in the Pleiades Open Star cluster, of course, only coincidently. This Pleiades open cluster is only about 400 light years away from our solar system.

Bobby McGehee

Figure 6.13. Pleiades Star Cluster. The most famous Open Star Cluster in the sky is the Pleiades. It is spectacular and can be seen without binoculars; It is also known as the Seven Sisters, and M-45, it is also the brightest and closest open star cluster. It contains over 2000 stars, is only about 400 light years away and it is only 13 light years across. Like other open clusters it revolves about the center of the galaxy within the galactic disk, as do most of the galaxy's other stars. Open clusters, like the Pleiades, form out of nebula as is currently occurring in the Orion Nebula.

Open clusters are interesting, but even more interesting are Globular Star Clusters. These star groupings have a completely different history and they are different in many ways. Globulars not only are usually much larger than open clusters and are spherical in overall shape, but they revolve (mostly outside) around and about our galaxy near the perimeter at various oblique angles to the galactic disk. Globular clusters, like their host galaxies, are direct descendants from the products of the deflagration front (Stage II). Proto-globular clusters, are from fragmented large gas and dust clouds, and probably start their formation at the same time as galaxies. Open clusters are products from supernovae nebula. Some say Globulars form from the halo around the proto-galaxy. However, in proto-globular cluster clouds, for reasons unknown, there is little large overall proto-globular cloud rotation, which makes them a periodic statistical anomaly. Without vortexing and related centrifugal forces, gas and dust cloud fragments accrete, contract, and evolve more rapidly into stars. Evidence indicates the required star formation time is about four billion years less without the proto-star cloud rotations than as occurs within pre-galaxy clouds. Astronomers who are specializing in acquiring knowledge about globulars are yet to explain much of the mechanical dynamics of motions for stars in globular clusters. Some suspect that dark matter gravity is playing a very significant roll. Globular clusters we observe today are of various sizes containing from 100,000 stars, to some containing millions. How some stars can oscillate back and forth through the highly populated cluster center without colliding more frequently, is a puzzling phenomena. However, there are a few globulars that exhibit what appears to be a black hole at their core; indicating there have been some collisions near their central core.

Also intriguing is the mystery as to how the proto-globular cluster clouds and the subsequent cluster assemblies lost their original linear velocity from the deflagration front without obtaining more rotation, while retaining their individual entity and geometry.

The slowing processs for Globular entities traversing through space is achieved through gravitational attraction trapping by the younger yet more rapidly decelerating, and at capture, slower larger massive proto-galactic clouds. Galaxies trap the faster but older proto-globular-clouds into orbiting, as they 'catch-up' to their slower and younger galactic proto-host. The glalxy and their captured orbiting globulars continue travels through space as a unit; Just as we observe for the MWG, and other nearby galaxies. This explains how that globular clusters are older than their host galaxies. The stars in globulars presumably ignite at about their same age as stars in galaxies; They just form at an earlier time compared to the stars in the younger, bigger, sibling proto-host galaxy.

Globular Cluster formation models are complex but are being developed by capable Cosmolgy-Physicist scientists. For the more inquiring minds, a text at the forefront of current globular knowledge is a book I find as very interesting. "Globular Clusters" was written and published in 1998 by Keith Ashman and Stephen Zepf. But remember, prior to now scientists did not have the NUT to help them understand the phenomena. I learned much from their text, learning things that I did not know that I did not know. It is fun to extrapolate and ponder form new knowledge.

Galaxies continue to lose some of their linear velocity as they are gravitationally annexed with other galaxies and become members of galactic clusters, subsequently revolving about and in their mutual gravity fields. Our Milky Way galaxy is in a family of mutually orbiting galaxies, called the "Local Group".. Globular clusters are observed around our two closest galaxies (referred to as The Magellan Clouds) as well as other galaxies. Some host as few as one, and others host as many as 48,000. Globulars. There are several theories as to why spiral galaxies are shaped as the are, and all theories may be at least partially correct for different galaxies.

Individual stars that form in the large proto-galaxy whirlpool vortexes (not limited to whirlpool appearing spiral galaxies), are later to accumulate the mass necessary for star ignitions, as evidenced by the much <u>younger age of our galaxy's star populations</u>, even though the galaxy is much larger than its' globular clusters. (The spiral arms just make the overall galaxy look like a whirlpool; the arms are not flowing inward to the center). Apparently the centrifugal forces within the individual star vortices are influenced by the galactic vortex, as evidenced by rotation direction uniformity. Rotation centrifugal forces delay and lengthen the time for gravitational contraction of the localized gas and dust cloud fragments into mass concentrations needed for individual star formations.

Starting with knowledge that all the mass in the vicinity of the Milky Way galaxy condensed from photons at approximately the same time. (All within less than a billion year). It is particularly interesting to note the oldest stars in the Milky Way galaxy are about 9 billion years of age, while the oldest stars (white dwarfs) in the galaxy orbiting M4 globular cluster are calculated to be at least ~13+ billion years old. A minimum difference of ~4 billion years; no small difference. This indicates that proto-star compacting processes are inhibited by centrifugal forces within the numerous and individual vortices, as well as by the overall galactic rotation in the proto-galaxy. It will be a surprise and a puzzle if the other 150 to 250 globular clusters around the Milky Way are found to be of different ages than M4.

Long before galactic star formation, proto-star cloud fragments are entrapped in the larger proto-galaxy vortex. Proto-star cloud fragments, become vortices as they contract from and within the much larger progenitor galactic cloud fragment. The nuclear furnaces of the individual stars randomly and spontaneously ignite as their mass grows and internal pressures increase through continued accretions, and they then start radiating light and other electromagnetic waves. If you have ever observed while flying over a city at dusk, the lights seem to come on in a random manner. The lights initially turn on within a globular or within a galaxy, in a similar random manner, of course with a different time scale. The proto-galaxies are feebly illuminated before their star ignitions begin, by the light from the stars that ignite about 4 billion years earlier, in globular clusters. It is not likely that we may someday be privileged to observe this.

One of the paradoxes solved by the NUT is to explain the inability of astronomers to find any of the hypothesized Population III stars. Steven Weinberg, Physics Professor, is the author of "The First Three Minutes", is recognized as one of the foremost knowledgeable persons of physical processes of the hypothesized BB. In his book he explains; At the end of the first three minutes (after the BB) the entire nuclear material content of the universe was 73% Hydrogen and 27% Helium. Nuclear physicists agree heavier elements could not have been generated by the BB. A hypothetical explanation by BB supporters for how the oldest observed stars, called 'Extreme Population II', contain some heavy elements; The hypothesis is that the early BB produced material was processed through some short lived, (only a few hundred million years), super-massive stars, referred to as Population III stars. These stars fused hydrogen and helium into heavier elements and then through supernova explosions, dispersed the new material into space to later coalesce into the Extreme Population II stars. The phantom Population III stars have never been found. Population III star fans seem to overlook that if such an early huge star existed the remnant supernova cores would have populated the universe with larger than ever detected super massive black holes! A good reason why these have never been observed may be that they never existed, ever. With the New Universe Theory, the heavier than Helium and Lithium elements as found in the observed older "extreme Population II" stars, are produced by the deflagration front in the Stage II velocity-enhanced-gravity era.

The most luminous observable objects in the universe are large massive galaxies and quasars, which were and are huge quantities of stars colliding and congregating near the core of a huge vortex as they disappear into one or more black holes near their galactic centers, destroying and devouring many hundreds of millions of stars. Bright novas often occur in the vicinity of the young galactic cores. Many fascinating and bizarre objects came into existence later, such as supernovae, neutron stars, and even quark stars which for the first time were only recently discovered, near the turn of the millennium.

(Fulvio Melia, Professor of Physics and Astronomy at Arizona University, recently (June 2003) published a book on "the black hole at the center of our milky way". His book is exceptionally intriguing and is recommended reading for all who are interested in cosmology. The central region of our galaxy is truly a mysterious and bizarre, but real place. Melia and other "center of our galaxy' researchers are modern explorers in the true sense of the word.)

The farther objects are from the center of the universe, the younger they are, and therefore currently still retain more of their initial high velocity. The objects closer to the center of the universe were formed early-on and having interacted with other objects over a longer time; have lost more of their initial linear velocity.... however, all are slowing asymptotically towards zero linear speed. It is estimated the first objects will have lost about 4% of their original linear velocity, which was the speed of light, by the time initial stars are generated and their light emission begins. This estimate is based on the fact that very few objects have been observed above about 96 % of light speed. In the near future, with the newer, more powerful interferometer telescopes, this observed distance will hopefully be extended to the initial photon emitting stars, projected to be moving at over 96%, possibly near 97 %, of light speed. It is expected that we may observe some of these first stars starting their emission of photons within a few decades. (Other than the already observed 'background radiation') Evidence indicates our closest distance to the deflagration front is in the direction of the Abell galactic clusters, but the front is _far_ beyond them. If we ever 'see' the initial annihilation 'light' it will be at very long wave lengths (it is part of the COBE and WMAP light) due to the Doppler shift from the photon's formations, occurring at light speed in the direction away from the universe's center, and also away from us. Some annihilation produced photons are emitted in rearward directions towards the center of the universe as not all are consumed in mass production. Those photons will produce some type of detectable virgin background radiation into the universe. Since we are within a few billion light years (maybe less than 5) of the center of the universe, the radiation should and does appear almost uniformly distributed with respect to general direction, at least within our current ability to detect. Spectral analyses of background radiation combined with Doppler analyses may someday precisely substantiate where we are located within the universe. The 2003 published data from WMAP (Wilkinson Microwave Ansitropy Project), reveals a few 'broad' bands of gamma wave background. However, discrete frequency data are needed for deflagration front spectral analyses.

The initial deflagration started at what is now the center of the universe. Starting from the speed of light, it takes a few (between 2 and 5), billion years of deceleration time for the processes of proto-matter conversion (reduction mechanism) to produce the first spontaneously igniting stars. The first star mass to be generated was precipitated at a distance of about 2.5 billion light years from that initiation site, initially moving outward at near the speed of light, but decelerating as it moved away. Therefore, the core of the central universe would initially have been a large void volume, as proto-matter is swept out in all directions by the deflagration front, as it transforms into mass. However, through super-galactic and 'structure' rotations, the initial void central core of 5 billion light year diameter three dimensional region may now be approaching similar density and may be almost representative of the nearby surrounding regions. An unexplained (to date) paradox that needs to be resolved is that we are at a distance of about 3.5 BLY (derived later in this document) from the center, which puts our region of the universe in or near the edge of the original core void? Red shift survey analyses do not yet reveal this. Someday soon, with further analysis of current and new infrared survey data, from present and new generations of telescopes, we may identify evidence for our universe's initial, first stating site. However, one explanation could be the above mentioned large scale circulation and 'fading away' vortexes may have erased the evidence.

When we are aiming our new technology telescopes in a direction away from the universe's center, we may soon observe the 'dusk' lights from new stars, (first in globular clusters, and then in galaxies.) We will then be able to better estimate the time and thickness of the Stage II transition zones. If we see stars with red shifts to 96% of the speed of light, and that is at the nearest distance to the deflagration front, say 21.0 billion light years, (the 21.0 figure is assumed only for illustration purposes). The Stage II transition zones will then be known to take <u>at least</u> 4% of the distance light traveled in the initial aging of the **Stage III** universe. It is expected the entropy growth rate (via clumping and vortexing) is much slower to develop among the earlier higher velocity particles. The distance/thickness estimate of the three concentric **Stage II** spherical shells is ~2+ billion light years to production of the first star ignitions.

The New Universe Theory Concept reveals, the region that includes our Milky Way galaxy is an estimated three and five tenths billions of light years (3.5 BLY) away from the universe's center. Also the universe (**Stages II + III**) is somewhere between about 34.4 and 68.8 billion light years in diameter. These numbers are much larger than conventional thinking, and only appear too large. Basis for these estimates are developed later in this document. The **Stage III**, potentially observable central universe, presently extends to about 95% of the total diameter, and dependent upon direction, between 15.5 to 39.4 billion light years from earth. (If the distance from us to the "dusk-lights" is measured in some particular direction to be 16.1 billion light years away, one year from now that distance will be 16.1 billion plus a growth of only .000000001%.

A new view of the universe's growth is illustrated and described in the section, "Size of the Universe". It is exciting, anticipating the new knowledge that will be revealed from additional deductive analysis of existing and new data, using the New Universe Theory Concept understanding and thinking. Including full compliance with the laws of physics. There is a tremendous amount of data available about the geometry and dynamics of the universe that can soon be re-analyzed in the light of the New Universe Theory Concept understanding. I will, from my 'amateur cosmologist's arm chair', be listening and watching anxiously for more data from professional astronomer's observations, and anxiously watching for new discoveries from old data as well as from new observations.

Deflagration sub-stages progress outward from, while trailing the preceding annihilation front as it travels outward in all directions at light speed. The NUT view of the universe's growth progress is illustrated in the section "Size of the Universe", but first it is helpful to comprehend if the 'busy chart', Figure 6.1, is reviewed. Information not included in that chart are the dimensions of Stages and Sub-stages. It is recommended that the Figure 6.1 chart be pondered while relating the dimensions listed in Figure 6.14 Table. At this time the dimensions are mostly surmised and presumed. Future mathematical modeling by particle physicists will develop more precise data to supercede the presumed dimensions in the table. If I were at the other end of my career, I would relish that modeling task.

N.U.T. Phase	Radial Dimension of Mechanism
Stage I	∞ ....Infinity......?
Stage II A	5.0 .... Centimeters
Stage II B	500 ....Centimeters
Stage II C	2.5 ....Billion Light Years (BLY)
Stage III	~ 20 + ...Billion Light Years

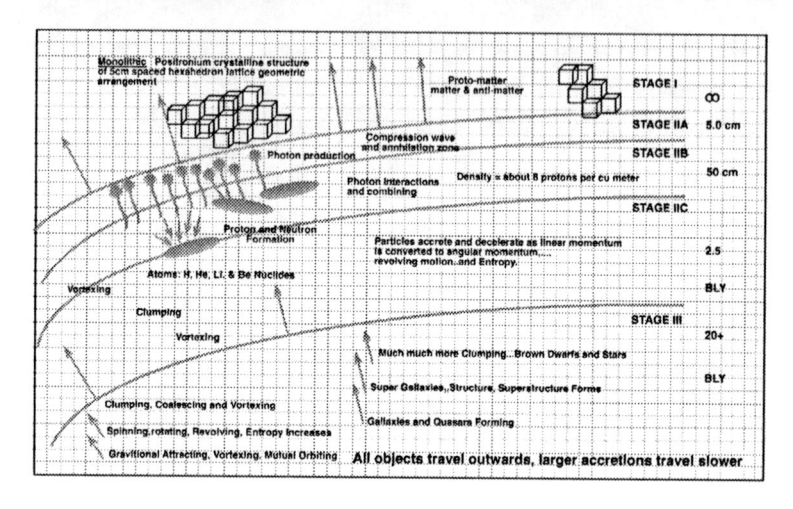

Figure 6.14. Table. Deflagration Distances. Compare the table data to that "busy" chart (Figure 6.1.). Stage II is the total thickness of Stages IIA, IIB, and IIC segments of the deflagration wave. The wave (all of Stage II) is traversing through the proto-matter (Stage I) and is leaving (dragging) behind the decelerating debris which evolves into the observable universe (Stage III). The wave thickness remains a fixed dimension as it continues to progress outwards, forward, and through more proto-matter. Stage II is 2.5 Billion light years thick. Now you can include in this figure the distances through which the reduction mechanisms function.

New Universe Theory, Concept Recap:

Stage I
Primordial Matter:
    Monolithic Crystalline Positroniums.

Stage II
Deflagration / Reduction Mechanisms:
    A: Annihilation Zone: Leptons (+ & -- electrons) to Photons.
    B: Photons Transforming to Hadrons (Neutrons & Protons)
    C: Hadrons to Nuclides & Clumps (Isotopes, to Brown Dwarfs, etc.)

Stage III
Observable Matter (Mass and Energy):
    The Universe.

Chapter 7 ... Entropy & Deceleration

What is "Entropy"?

Many think when the obscure term "entropy" is used, the user is throwing out a smoke screen. I have read many articles and books in which the term is used to describe a process that they themselves do not understand. But, when used in The New Universe Theory (NUT), which is the biggest example of thermodynamics in action, entropy is used properly and it is not confusing, misleading, or misunderstood. This chapter provides a general discussion on the subject "entropy" and then proceeds to explain the entropy influence on generation of the universe.

Webster's Encyclopedic Dictionary definition; *Entropy* is "a measure of the degree of disorder of a system". In thermodynamics it is a direct indication of unrecoverable energy. The second Law of Thermodynamics says that entropy in any system is forever increasing. The universe is the largest system of all.

In the NUT entropy is the deceleration impetus, but it is not a force.

The universe is made up of two known constituents; Space and Matter. The New Universe Theory includes a concept for primordial matter and for the processes through which primordial matter evolves into the universe's existing forms of matter. All of the processes comply with the proven Laws of Physics; any theories that don't comply presumes the laws are or were not always valid. But the laws, to be laws, have been proven, and do not allow for exceptions, therefore the errant theories are invalid. There is no evidence that the Laws were ever different. The only arguments with the laws, that I have ever heard, have been to allow the BB to stand. These laws are enforced by nature, not just because somebody or a committee says so.

<u>Matter</u> cannot be created or destroyed and is made up exclusively of two components: Mass <u>and</u> Energy; either of which can be transformed into the other, and each has many different forms. <u>Mass</u> is made up of basic elementary particles, which, when combined in various arrangements and proportions, make up all of the elements and compounds for all of the many physical substances. <u>Energy</u> is in several forms, each of which can be converted into the other forms; Mechanical, Electrical, Chemical, Magnetic, Heat, Electromagnetic. Energy can be either kinetic or potential; Kinetic is active, and Potential is static. <u>Entropy</u> is a form of energy, and energy is a form of matter which cannot be either created or destroyed. Entropy is unavailable for conversion into other forms, and increases at various rates from almost all processes. Since it cannot be converted to other forms, it continues to increase throughout the universe; which is the second law of thermodynamics.

Entropy, contrary to the implication by the Dictionary definition, cannot be directly measured. Entropy can be explained by fluid (air) flow examples.

Example 1: In this example we start with a straight continuous area pipe about four inches in diameter. If we measure the internal air flow pressure, temperature, and velocity 20 diameters upstream of a position in the pipe and also 20 diameters downstream of the same position, we will find the flow conditions are identical, consistent with the continuity laws (no air or heat is added or taken away). Now, if inside the pipe at a position half way between the measurement stations, we place a short flow area reduction, such as a thin flat plate with a small 1"diameter hole at the center. The two pressure measurements will now be different due to dynamic flow losses through the obstruction as a result of contraction and re-expansion and vortexing induced into the flowing fluid. This measured difference can be predicted if the flow coefficient of the obstruction is known. Engineers have cataloged 'Flow coefficients' of various shapes of flow obstructions. These 'coefficients' are usually empirical (determined from experiment). The dynamic pressure loss is to entropy. The entropy energy is non-recoverable.

Example 2: Another more familiar example is the outflow of air from an automobile tire when the valve core is depressed so that the valve is opened. The temperatures of the air inside the tire, and of the out-flowing air, immediately outside the valve, are equal. There was work (energy) required to compress the air into the tire, but since air is now released and expands into the atmosphere without doing any work, this released potential energy produces only air flow and its velocity decelerates as it dissipates into vortexing and mixing with the ambient air; energy is thereby transformed into entropy.

Entropy, consistent with the second law of thermodynamics, increases in any (and all) system(s), towards a maximum. As previously stated, "Hubble numbers" that have been calculated from red-shift observations are not the universe's rate of expansion, they only indicate the growth rate of entropy. The New Universe Theory proto-space and proto-matter starts at zero entropy (complete orderliness). After proto-matter positronium annihilations, the photons coalesce and precipitate into mass particles with linear momentums. The momentums are partially transferred to entropy through mixing and vortexing. Numerous repeated particle interactions are unavoidable, and the linear velocities of mass particles continuously decrease. Entropy growth from the momentum of the fluid flow of new mass objects in space starts immediately, from initial mass formation, and then continues on and on; to borrow a phrase from Carl Sagan, continues for "billiouuns and billiouuns" of years, to ever increasing levels of entropy. (We all continue to respect and admire Carl; I wish he were still here to present a TV series on this New Universe Theory.)

Forms of angular momentums:

Revolve *is to move in an orbit.*

Rotate *is to turn or cause to turn about an axis.*

Wherever we look, we see angular (rotational and revolving) momentum. Lots of it. The earth is rotating, as is the moon, and all of the moons and other planets. Our sun is rotating and so is our entire solar system. All of the other star systems are also rotating. And then all of these stars systems make up our Milky Way Galaxy, which is rotating. Our galaxy is revolving as part of the super-galaxy known as the Local Group, which is also rotating. In fact, when our earth telescopes view the depths of space, we literally see billions and billions of galaxies. All are rotating and revolving, individually and in combinations. All of this angular momentum is energy. Momentum is quantified as the product of mass multiplied by velocity. That means a lot of mass is rotating and revolving 'out there' and this energy came from somewhere. This direct observation is available to everyone. According to *the NUT, all of this angular momentum was originally in the form of linear momentum. All elementary mass particles were originally traveling with very high linear velocity (momentum) immediately when they transformed from photons. Through interactions these angular momentums came from the original linear momentum. The mass objects of all sizes in the universe are not traversing at anywhere near the original speed of light as the objects separate from the front. However, at great distances, (closer to the front) and as the front progresses outward, we observe the objects are still at very high velocities. They too will slow their linear velocities as mixing and rotations increase in their region. Another Law of Physics is that momentum must be conserved. But it can be transferred. All of this observed motion strongly supports the New Universe Theory. The BB theory has no explanation for the angular momentums or the higher velocities with distance as envisioned by those who give credence to that theory. The NUT envisioned decrease in linear velocity is what in thermodynamics is recognized as increasing **entropy**.*

Recalling the **Laws of Physics**, Newton's second law of motion states that any change in an object's velocity, either an increase or a decrease, requires the application of force. Newton's Laws of Force and Gravity have been proven and are both credible and unavoidable. Yet, there are no acceleration or deceleration dark energy 'forces' to act over the huge distances from the outer reaches of the universe to the central regions of the universe. The riddle is why are objects, throughout the universe changing velocity (decelerating)? The answer has already been discussed, but is obscured by the laws for conservation of momentum that state that the energy of momentum cannot be destroyed. ....or created. ..... But energy can be transferred to energy in other forms, as in the case of deceleration of astronomical objects, energy is transferred from linear momentums to angular and rotational momentums. What are the deceleration rates? Entropy, with respect to linear momentum, produces an equivalent end result as would occur by 'deceleration forces', but only to the forward vector component of velocity. The Hubble numbers do not directly define deceleration rates, but do indicate local aggregate reductions in forward velocity.

Objects with momentum (the product of velocity and mass) occasionally pass near enough to other objects that their mutual gravitational attractions trap them into mutual orbiting. Part of the initial linear velocities are converted into angular velocities with each of the objects' total, tangential and linear, velocity being equal to its original linear velocity. Once in orbit, the tangential velocity on the opposite side of the orbit is in the opposite direction, and thereby the object's transitional velocity is reduced, while it still retains all of its original momentum energy. Conservation of overall energy is preserved while linear velocity is reduced. This process occurs many times over and the results act like deceleration forces. Where millions upon millions of particles are involved, the gross movement is known as fluid flow. The flow resistance produces random motion, spinning and revolving about one another within the fluid, much of which becomes an increase in entropy. Particles clump, through collisions and accretions, and vortices absorb velocity in one direction by conversion to changes in travel direction, while conserving energy. Overall deceleration continues to occur while the system remains consistent with the Law of Conservation of energy and it simultaneously stays consistent with Newton's Laws of Motion.

When the famous World War II five star General Douglas MacArthur, made that famous statement in his retirement speech, I knew immediately that he had studied thermodynamics and fluid flow. He borrowed and modified the statement which was originally made by Alan Shapiro, who is considered as the Father of Modern Thermodynamics. (I have found by searching the internet *that there are at least nine individuals credited as Father of Thermodynamics). Shapiro's original quote was "<u>Old Vortices never die, ... They just fade away</u>". Shapiro's statement is certainly valid in his field of fluid flow where vortices eventually dissipate into micro-vortices and then into random molecular motions. However in the large rarefied expanse of outer space with no boundaries, small vortices coalesce into larger, and then into still larger vortices; Simultaneously, the particles gravitationally coalesce and converge (clump) with other clumps and vortices contract to form rotating stars, binary star systems, galaxies, galactic clusters, and eventually, large slowly rotating structures. (<u>Most</u> of the stars that are observed in the universe are members of binary, tertiary, quadruplet, or other multiple star systems.)*

What an interesting, informative round table discussion we could have if V. M. Slipher, Milton Humason, Edwin Hubble and others such as Fred Hoyle were alive today to discuss my New Universe Theory. V. M. Slipher of Lowell Observatory was the first to discover both Andromeda is a galaxy not unlike the Milky Way. Also, he is the discoverer producer of the data showing increasing red-shift with distance that was analyzed and verified a decade later by Humason and Hubble, and then declared by Hubble to be "Hubble Law". Humason, a humble self educated individual actually was the one knowledgeable in line spectra to do the analysis, and I have never read of any resentment on his part for Hubble's bold claim related to Slipher's discovery. I believe Humason and Slipher would now chuckle and crack a slight smile of vindication, as my New Universe Theory shows the 'Hubble Law' is the only part of the discovery that is not valid. Certainly Hubble is to be credited with making the public aware of progress made by the scientific community. Also many scientists began to ponder and question previous declarations and continue to search for more definitive explanations as to why is the universe as it is? I have lost count of observed puzzling phenomena for which the New Universe Theory provides an answer that otherwise required deviation from the Laws of Physics.

I encourage and personally welcome discussions within the framework of the Laws of Physics. It is my intent to provide an internet web site to receive and present continuing suggestions and responses.

The universe's entropy change rates, at first, appear as linear functions, but have only been compared to other distance candles over limited distance ranges. Decelerations throughout both the NUT Stage II and early Stage III mass processes are significantly different from linear, as the environment that allows mass accretions and accumulations, varies significantly. Velocity enhanced gravitational interaction, and particle proximity interactions encourage and cause accretions and turbulent mixing. The data being gathered by the several sky surveys will provide non-linear information when correlated with other distance candles, will provide mapping to define the distribution of Stage III entropy growth rates. Entropy growth rates throughout the Stage II Reduction mechanisms and Stage III can then be defined by mathematical and computer models. Such computer models are yet to be developed. In the meantime this document uses estimates from graphical analysis. Much more research is yet to be done than meets the eye to correlate the red shift data with other 'same direction' distance candles for entropy mapping all of Stage III. Astronomers have much exciting work to do.

This next order of business should be to complete the red-shift maps of the cosmos which is a high priority need for a comprehensive understanding of the universe and the dynamics of cosmology. Analytical processing of the red-shift cartographic data when correlated with other directly applicable distance and direction candle data, will expose revelations that here-to-fore are yet to be imagined. Early-on significant red-shift surveys were ambitious projects were by research teams of Margaret Geller, John Huchra, M. Kurtz, V. deLapparent and others. Follow on projects included Mark Davis, Dave Latham, and John Tonry. As they progressed in their CFA and CFA2 survey projects.

Two early sky maps are referred to by free lance cosmology writer Govert Schilling as pioneering pie slices. The Center for Astrophysics Red-Shift Survey (CfA) uncovered the Stick Figure and the Great Wall of galaxies running through the Coma Cluster; (these maps span about 1 billion light-years). The map on the right is the Las Campanas Red-shift Survey, with its 26,000 targets (spans about 2 billion light years), and it shows that such structures were commonplace. (Maps courtesies of John Huchra, Harvard-Smithsonian, CfA, and Stephen D. Landy, College of William & Mary). As a direct response to my specific question at an Astronomy Club meeting, (Guest speakers were Margaret Geller and John Huchra), Ms. Geller stated (in 1999) their project "will require until about the year 2020" to complete. Since this team, including Huchura and Geller (Path-finders/Trail-blazers of the cosmos) started their project, technology has advanced significantly, and now, several ambitious sky surveys have since been initiated, which with new technological equipment accelerates by many fold the rate of obtaining red-shift data. The February 2003 Sky & Telescope magazine article by Govert Schilling (ref 43), stated some of them cover large fractions of the sky, while others probe tiny patches but reach out to very high red-shifts.

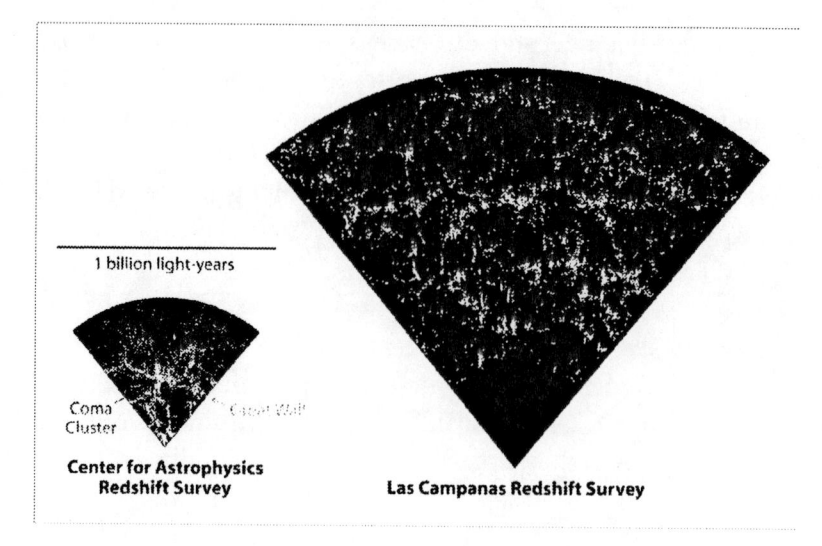

Figure 7.1. Stick Figure A stick figure emerged as Astronomers Margaret Geller, John Huchra, and team began plotting their data from three six degree thick slices of their CfA survey, which was approximately centered on the Coma Cluster. As additional data was plotted the stick figure faded into the mix of other data. With this survey they also discovered the array of galactic clustering which has become known as the Great Wall

Figure 7.2. Structures and walls As data from the CfA survey was compiled, many super structures start to emerge. This view includes six of the six degree survey slices. Voids, strings and walls of galaxies and many super-galaxies come into view.

John Huchra also uses the analogy of a hockey puck to describe and illustrate the distribution of observed galaxies. The radii of the circumferential rings drawn in Huchra's "puck" represent speeds of 5,000; 10,000; and 15,000 km/sec. Ironically, the largest circle corresponds to only 5% of the speed of light. To provide a comprehensive view, data is needed as far away as 95 to 97% of the speed of light (19 times the coverage volume of the Huchra puck). What we would like to see is the mixing and entropy growth all the way to the vicinity of the location where matter flows out of the Stage II mechanisms, into our Stage III universe. In this puck view, we are looking far beyond data from the information producing generations of Slipher, Humason and Hubble, but we are still only viewing nearby galactic distributions.

Bobby McGehee

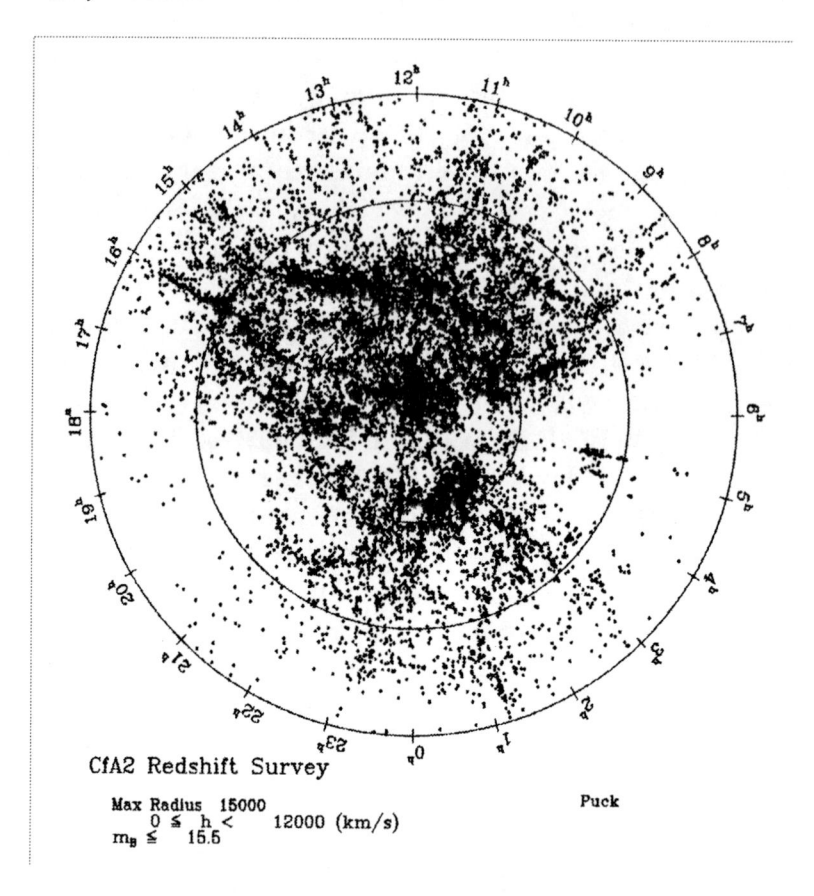

CfA2 Redshift Survey

Max Radius 15000
$0 \leq h < 12000$ (km/s)
$m_B \leq 15.5$

Puck

Figure 7.3. Puck. A 360 degree plot of red shift data out to a sky velocity depth to 12,000 Km/sec is a cylinder of data in a volume, as Huchra refers to as shaped like a huge hockey puck above the equatorial plane. The survey to acquire this data was conducted by Huchra, Geller, and others from the CfA2 project. This figure is a reproduction of the chart from John Huchra's web site: [http://cfa-www.harvard.edu/~huchra]

Proportionately, on the same scale, to enclose 95% of the diameter of the universe, like the 5% enclosed by Huchra's 'puck', the volume would be increased to the diameter of a 50 gallon oil drum. Only 5% is represented by a 3 inch diameter puck! However, Huchra's symbolic 'puck' is a novel and good way to convey the shape of the volume measured by his studies, and gives us a comprehensive size image. The data illustrates a major accomplishment, but it is only a start. To date, the total portion of the sky, in terms of volume, mapped by all survey projects is estimated at less than 1.0% of the universe's volume of many billion-cubic-light-years. (based on Universe size from Chapter 9 and 10 graphics). Eventually, we should be able to map almost halfway to the deflagration front except for the regions obscured by other objects including our own Milky Way galaxy. The estimated total map-able portion of the universe is about 10% of all space inside the universe. Light from the outer half (radius) of the universe has not yet had time to reach us. From volume difference equations for concentric spheres, the out of sight range, (outer half of the distance), contains seven eighths (87.5 %) of the universe's volume and all matter contained therein.

Govert Schilling, in February 2004 Sky & Telescope, (ref 45), presented an updated map of some additional data from the Sloan Digital Sky Survey with the CfA2 surveys. These maps are similar to those in Figures 7.2 and 7.3 and show more of what some have referred to as 'structures'. These galaxy groupings are appropriately described by the project astronomers as 'chance arrangements', and furthermore, they are only temporary. The continuing rotations and revolving super galaxies produce short lived alignments some of which have been given the mis-nomer of structures and walls. When these surveys are taken again in a few million years, the alignments will be differently arranged.!

Modern Red-Shift Surveys				
Survey	Descriptive Title	Targets	Red-Shift
SDSS	Sloan Digital Sky Survey	600,000	18
2dF	Anglo-Australian 2 degree field	225,000	19
6dF	Anglo-Australian 6 degree field	100,000	12.8
VIRMOS S	Visible Multi-Object Survey	100,000	22.5
VIRMOS		50,000	24.5
VIRMOS UD		13,500	26
DEEP I	Deep Extra-galactic Evolutionary Probe	1,100	24
DEEP II	NASA & NSF	10,500	23.5
DEEP II-D		50,000	26
IRAS 1.2 Jy	Infra-Red Astronomical Satellite	5,600	1.2
IRAS/PCS		15,000	0.6
Las Campanas	New Mexico	21,000	18.2
CfA1	Center for Astronomy Survey	2,400	14.5
CfA2	CfA S (Harvard)	15,500	18
SSRS1	Extension of CfA1 and CfA2	2,030	1.2
SSRS2	Second Extension	5,400	15.5
2MASS	2 Mic. All Sky Survey, <11.25 z	24,000	11.25
2MASS	z = <12.2	100,000	12.2
WMAP	Wilkinson Microwave Ansotropy	NA	NA
COBE	NASA	NA	NA

Figure 7.4. Table Modern Red-Shift Surveys. Data for the listed Sky Surveys table was mostly provided by Govert Schilling, some of which he said was supplied to him by John Huchra and several others. This is not a complete list and hopefully this list will be much longer in the near future.

Thanks to several ambitious sky surveys, maps are now available of velocity of many specific objects at their specific directional locations. The object's true distances are not known, because directional unique distance candles have not yet been catalogued for correlation with velocity and direction. Astronomers in the past have assumed that there is a single Hubble number that applies in all directions. (If the Big Bang was valid, one universal Hubble number would appear to be appropriate.) In the future, after the sky has been calibrated for Hubble number verses direction, we will be able to produce entropy maps, and then accurately define where we are in the universe.

Since the suggested surveys have not yet been accomplished, a cursory analysis allows us to estimate our location in the universe. *First*, as previously stated we tentatively assume all of the Hubble numbers are valid, just acquired in different directions. (We accept this with reservation, knowing that the distance candles were probably measured in various other directions.) *Second*, we plot the largest and smallest Hubble numbers of modern times on a graph of velocity (y-axis) verses distance (x-axis). these 'Hubble lines' are then extrapolated to where they intersect 96-97% of the speed of light, which corresponds to the trailing edge of the outward progressing deflagration front. This is also the periphery of the universe. If the longest and the shortest are assumed to have been measured in opposite directions, that is, approximately 180 degrees apart, we can add their distances to the deflagration front. This total distance describes the initial NUT estimate for the diameter of the Stage III universe.

[Figure: Distance vs. Velocity Graph of Abell Clusters; y = 117.67x − 4728.3; x-axis: Distance in MegaParsecs (0–800); y-axis: Velocity in km/s (0–140000)]

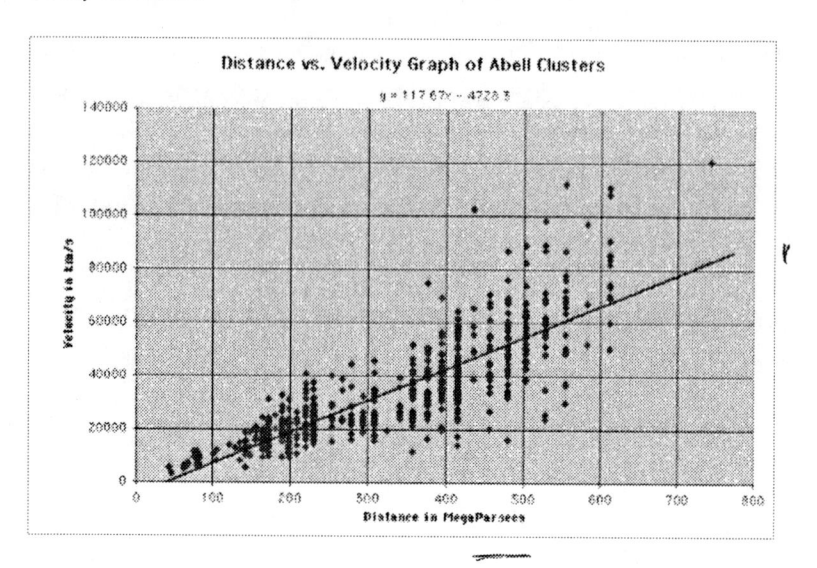

Figure 7.5. Abell Clusters. The three person team that studied this huge galactic cluster produced the highest Hubble number measured using modern technology. Their data and calculations resulted in the Hubble number of 117 which is used in the NUT for determining our closest proximity to the deflagration wave. The distance to the Abell clusters range between one-half to two-and-a-half BLY (~75 to 750 Mega-Parsecs) from the Milky Way Galaxy. *The team was Gibis, Choi, and Mittappali. This chart is reproduced here with permission. (contributors home pages are: emgebis@amsi. edu ; ramrom@imsa.edu ; and mchoi@imsa.edu.). Their calculation methods are included in Appendix II.*

The next chart (Figure 7.6) is a plot of the highest and lowest measured Hubble numbers of recent decades. The steepest slope has the value of 117 by Astronomers Michael Choi, Liz Gibis, and Sreeram Mittappali (measured in year ~2000); The lowest recent slope Hubble number is 50, by Tammann and Sandage (measured in year 1999). In this chart, the Hubble numbers (more appropriately called gradient lines) are extrapolated to 96% the speed of light which as shown in the following chart is also the deflagration front location and speed. The abscissa value is the base distance from the "Y" axis. The graph identifies the distance to the deflagration front from the observation site.

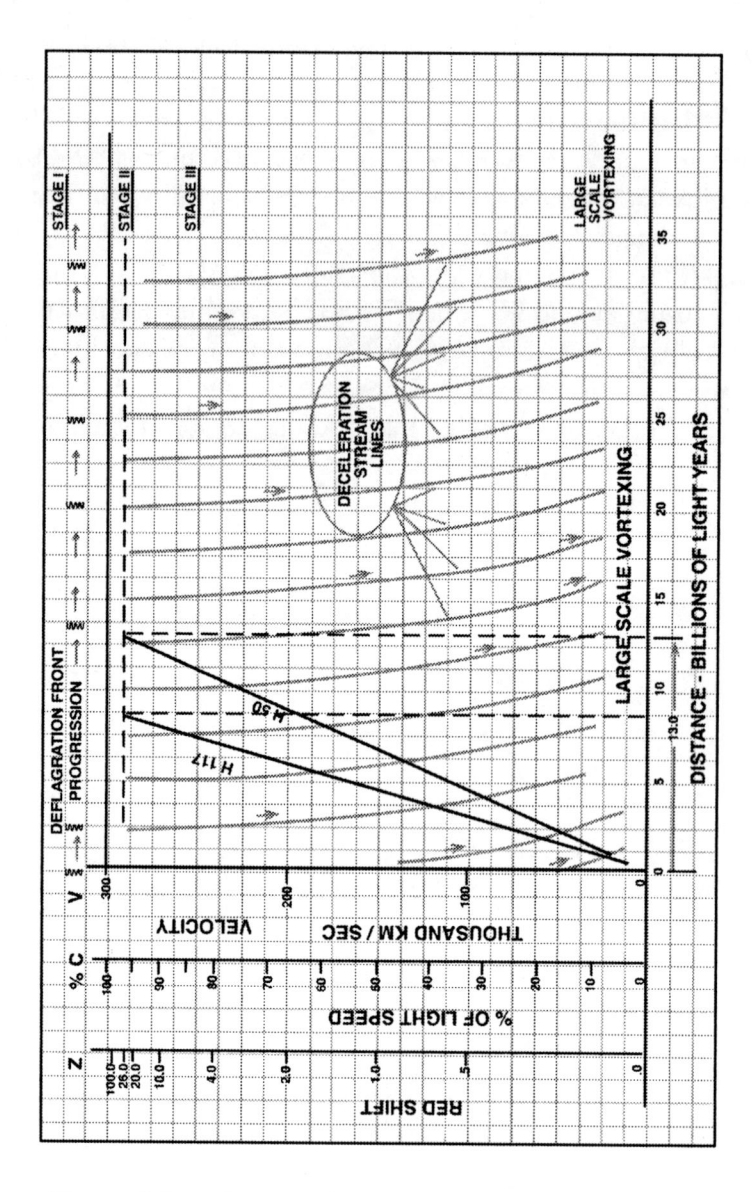

Figure 7.6. Hubble numbers indicate velocity gradients. Graphing the largest and smallest Hubble gradients on a chart of velocity vs distance, and extrapolating to the speed of light, then projecting back on to the distance scale reveals the farthest and the closest distances to the Deflagration front. Drawing the other Hubble Number direction/distances to the deflagration front provides the distance to the front in their directions (see Figure 7.7).

In the next graph, (Figure 7.7), the largest and smallest Hubble number distance projections obtained from Figure 7.6 have been drawn in opposite directions. This assumes the steepest (largest Hubble number) is towards the closest perimeter of the universe, and the lowest slope is towards the farthest point on the perimeter, as determined from the (Figure 7.6.) graph. Intermediate distance-to-the-perimeter hubble-number-lines were scribed to the semicircle-assumed perimeter of half of the universe. When these lines are in place, the measured angle should correlate with the direction in which the hubble number data was observed. (I do not have access to the data that is necessary to substantiate the respective astronomer's measurement angles). It also must be recognized that several assumptions are made to even make it possible to demonstrate this first-cut preliminary graphical analysis to reveal the size of the universe and where in it we are located.

The differences among most published Hubble numbers are explainable as due to the direction and distance over which the numbers were measured from the Milky Way's non-central position. Hubble numbers are directional-unique, and to be valid, plus something not previously understood, must be acquired from red shift data that is obtained in the same direction and approximate distance as the reference distance candle used for calculations.

The following are four tentative 'bold' assumptions made to allow this preliminary and tentative analysis to proceed. First, the Hubble numbers are tentatively assumed to be linear all the way to the deflagration front (also the speed of light); Second, for lack of any other data for direction vs Hubble Number (H#), angles in the table have been assumed to be in a plane in one half of the celestial sphere, (all within 180 degrees). Third, the largest H# (117) was assigned the angle of 180 degrees and the smallest H# (50) was assigned the angle of 0 degrees. (The NUT depicts these two are valid H#s as they do not require angular corrections for the velocity calculations. Explanations in Chapter 12). <u>Fourth</u>, the surface of the deflagration shell (periphery of the universe) is assumed to be spherical.

Other Hubble numbers in the table were plotted and extended to the inside spherical surface of the deflagration front (~96% of light speed). The projected distance dimension for the intermediate size Hubble numbers were then drawn from the origin to the periphery. Angles were then measured only to indicate from which direction the observers probably obtained their red shift data and to stimulate pondering. The assumptions and projections, combined with the known data, provide us with the first tentative cross section view of the universe.

Analyses of recent and yet to be acquired infra-red spectral data, can provide three dimensional maps of entropy at all distances and directions surveyed. Such analyses will then tell us much about the past and present growth of the universe, as well as more accurately telling us where we are located in the universe, including our distance from the center. But for now, we must work with assumptions and observational data that is available. That is what is presented in the graph.

The distance projections (Figure 7.6) from the Hubble number lines are towards the location where the extrapolated lines intersect the circumference of the universe (Figure 7.7) when scribed as a radius from their measurement location (Earth). It should be remembered that the universe is three-dimensional and our view is of a two dimensional cross sectional view. The Hubble number lines now located on the diagram of the universe are lacking the data showing the angle in the third dimension, but knowing that angle will not change our view of the size of the universe, assuming it is spherical. The actual data acquisition angle (specific red shifted object and the reference distance object) can of course be determined by consulting with the observer that acquired the data, or researching his or her documented records. (It is interesting to note, the Abell clusters are in the range of about .8 to about 2.8 billion light years away. Most Hubble numbers, if not all others, were from very much shorter range data. Determining the relative angles for the third dimension, and substantiating the other two will be tedious and challenging, but will be curiosity satisfying, interesting, and exciting.

The longest distance and the shortest distances to the deflagration front when placed end to end, tentatively describe the diameter of the universe. Our location is at the point of contact (origin) of the two vectors. Since the deflagration front travels at the speed of light in all directions, the center of the universe is located at the mid point of the sum of the two vectors. A line (diameter) of the universe goes through the center of the universe, and through Enid, Oklahoma!. *(My home town, a little levity here. On the universe's scale, anywhere on earth, the center of the earth, solar system, Rome, and even the whole Milky Way Galaxy are within the same minute point).* This tentative graphical drawing indicates that we are all located at a nondescript point about 2.5 Billion Light years from the origin and center of the universe! This graph and the dimensions it displays are initial, tentative, and these data are reconciled and adjusted to the 'real world' in the next chapter.

The initial tentative view of the universe shown here has yet to be reconciled with other knowledge, to more correctly reveal the universe's size and age.

Bobby McGehee

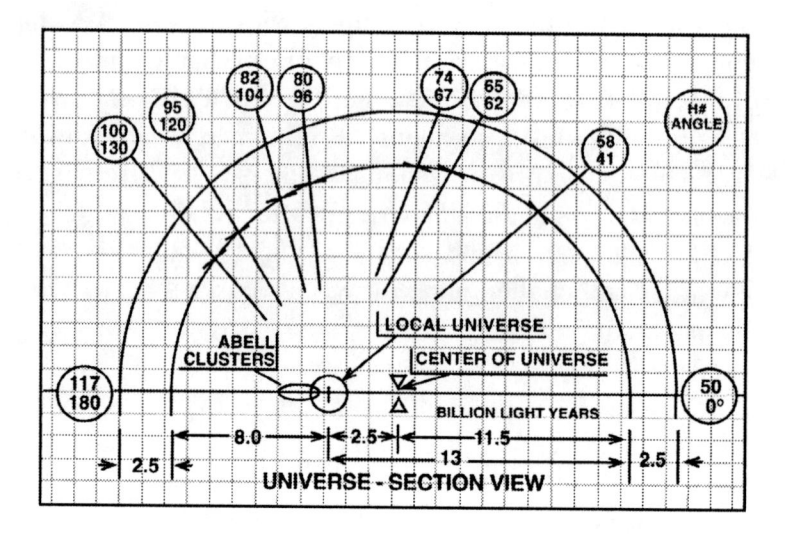

Figure 7.7. Universe Cross-section view. Placing the largest and smallest Hubble vector lengths determined from Figure 7.6, end to end, assuming they were measured in near opposite directions, reveals from this nomograph the hypothesized "diameter of the universe. Center of the universe is equidistant from the front in all directions, which allows us to deduct our distance (~2.5 BLY) from the center. Drawing the other Hubble Number direction/distances (using Figure 7.6.) to the deflagration front provides the distance from us to the front in their directions. When these distances are scribed as radii on the semicircle, hypothesized Hubble number acquisition directions are revealed. The other polar coordinate angle of measurement could be at any angle, for that Hubble number.

NEW UNIVERSE THEORY WITH THE LAWS OF PHYSICS

You have just glimpsed the 'first-cut' universe cross section view from the New Universe Theory which reveals new information here-to-fore unavailable, all by analysis and deductions from existing data that has been staring us in the face for decades. This graph shows where we are in the universe and it is easy to see our relative position to the center of the universe.

Hubble Number Direction from Nomograph					
Angle degrees	*H # Km/sec /Mega parsec*	*Hubble # Km/sec/ BLY*	*Astronomer(s)*	*Year*	
180	117	.003814	Choi, Gegis, Mittapalli	2000
130	100	.003260	de Vancouleurs, et.al.	1979
120	95	.003097	Aaronson	1980
104	82	.002673	Aaronson and Mould	1983
96	80	.002608	Tully and Fisher	1977
67	74	.002360	Edward Ajhar	2002
62	65	.002119	Mould, et.al.	1980
41	58	.001826	Allan Santage	1999
0	50	.001630	Tammann and Sandage	1982

Figure 7.8 Table. Hubble measured directions by nomograph. The variation and range of measured Hubble numbers over the time since Hubble numbers were defined is over a period of 70+ years. They are in units of Km/Sec/Mega parsec, and are also listed by Km/Sec/BLY. All H numbers are between 50 and 117. These data have been obtained from various publications by various authors and have not been verified with the indicated observer's records. As previously stated, the basic Hubble number data are thought to be dissimilar because they were simply measured at various unknown directions and at different distances, they therefore are towards different portions of the deflagration shell. The individual measurements are also thought to have been made by competent astronomers with reasonably good equipment.

With these assumptions, all of these numbers are accepted as valid within some reasonable measurement accuracy and equipment tolerances. The angles of measurement are yet to be verified with archived data.

Many of us are anxiously awaiting these types of graphic analyses from more refined and more complete data catalogued by directions correlated with distance candles, in three dimensions. The Sloan project (SDSS) achieved a significant start of obtaining these data, and now it needs to be analyzed for the size of the universe. Professor Mageon of the University of Washington at Seattle was project manager from early on, and may now decide to delay his expressed desired retirement. The project was and is a resounding success. But analyzing and obtaining more of these data in the light of the NUT perspective are surely too appealing for him to resist the call of Cosmology. I only know him from his TV presentation, and he appears too young, intelligent, and energetic to retire in the face of the challenges exposed by New Universe Theory. Results of these data analyses will be fascinating and so revealing of new knowledge. All of us old retired armchair cosmologists and amateur astronomers will have to be careful to avoid over excitement from pending discoveries.

Please recognize that the results in the first-cut cross section view of the universe are cursory and preliminary; this chart must be, and will be reconciled with other facts relative to the age of our region of the universe. The first reconciled view is presented in the following Chapters on the "Age and Size of the Universe". However, before we can go there, we must further examine the universe generation mechanisms. Graphically extracted using hypothesized data acquisition directions are listed in Table 7.8.

This clear perspective of the universe is now available because we have now opened our eyes and are now looking at the red shifts, recognizing these red shifts are indications of individual light source velocities <u>relative to us, and nothing else</u>. The Hubble lines are deceleration gradient lines and the decelerations are due to the thermodynamics phenomena, "ENTROPY".

Chapter 8 ... Fluid Flow

For those that have had the fun of studying Thermodynamics, which is part of almost all Engineering fields; Aerodynamics, Petroleum, Mechanical, Civil, Hydraulics, Nuclear, etc., the universe development from the deflagration front has probably already come to mind as a fluid flow analogy. However, <u>Fluid flow is not just an analogy for development of the universe. Our universe is in fact being produced through a series of fluid flow processes.</u> Fluid flow is discussed here not only because it helps one understand the New Universe Theory, but also since the universe is developing through these processes, and Fluid Flow will necessarily be part of the framework that will be used for computer modeling of the NUT. An in-depth technical understanding of fluid flow is not necessary to comprehend basic fluid flow processes that relate to the Universe's development from the deflagration front.

For those who may not be familiar with fluid flow dynamics; Here is a cursory short course in a few paragraphs and with a couple of photographs.

All fluid flows are similar in character as they progress. Fluid flow from a plenum source starts out as a <u>laminar flow</u>, which is a short lived smooth tranquil process. In our New Universe Theory, the hadron (proton and neutron) flow from the photon precipitation "pool" is a fluid flow of special sort. ...at extremely low density, and at the highest possible velocity. In conventional fluid flow that is familiar to engineers there is always a boundary, (except in the vacuum of space) and at the boundary interface there is friction which is a resistance to flow movement. As the flow continues some turbulence and vortices always develop in the flow friction shear area near the boundary. It has been determined experimentally that when fluid flows a certain distance the fluid is said to have reached the lower critical, which is where turbulence starts to propagate from the boundary shear area into the fluid. The distance from the lower critical to <u>fully developed</u> turbulent flow is known as the <u>transition zone</u> and it is usually over just a very short flow distance.

At the distance where turbulence has propagated throughout, it is said to have reached the upper flow critical, where the flow is referred to as fully developed. Engineers have a convenient term for calculating where these criticals occur. The term is called Reynold's Number, and it is calculated as the product of velocity, density of the fluid, and distance traveled, all divided by the viscosity. <u>Reynolds number is</u> in essence <u>a ratio of dynamic forces to viscous forces</u>. Viscosity is the flow resistance offered by a fluid relative to motion of its parts. Reynold's Number is purely empirical, as are the values of viscosity. (Empirical means derived at least partially, if not mostly, from experiment).

Flows in all fluids are unique, yet <u>all occur in the same basic manner,</u> which is clearly visible in a simple smoke plume. Smoke is a mixture of minute particles suspended in a heated gas. The particles provide 'tracers' for flow visualization. Flow progresses upward, and so does the description adjacent to the picture.. There are always three flow zones: Laminar, Transition, & Turbulent.

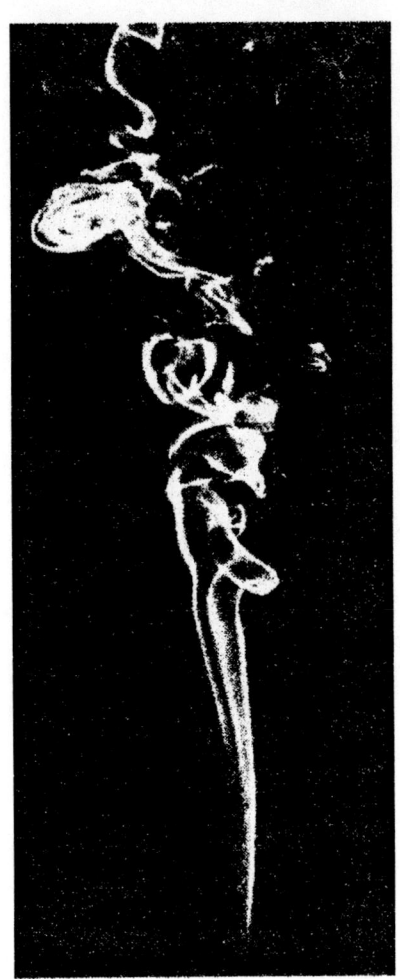

"Turbulent" fluid flow becomes dominant as flow progresses. The vortexes become larger as they merge. Vortex cells build into larger structures, but some become separate entities.

"Transition" is where the flow becomes non-laminar and small elemental flow regions start having reverse flow velocity components.

"Laminar" is the term to describe initial smooth flow, but viscous effects start slowing the flow at the layer interfaces. Dynamic stress has not yet caused enough localized drag to achieve reverse velocity vectors.

Figure 8.1. Smoke Plume. Flows in all forms of fluid occur in the same basic phases. The smoke particles in this flow of heated gas, makes the three basic phases clearly visible; 1) flow starting as laminar, 2) the transition from laminar to turbulent, and 3) the turbulent flow phase. Republished by permission Krieger Publishing Company, from Introduction to Fluid Mechanics, Stephen Whitaker, 1992.

Bobby McGehee

Flow phenomena in the three phases is also visible in a fresh water stream (liquid fluid) like in this Idaho creek. The photograph and explanatory caption identifies all three phases.

NEW UNIVERSE THEORY WITH THE LAWS OF PHYSICS

Figure 8.2. Liquid Flow in Nature. Flow phases in liquid fluids are observable in this Idaho fresh water stream. This stream is flowing from left to right; It flows from a wider pool just left of the photo where the water is placid, clear and laminar flowing. As flow enters the field of view the stream is becoming opaque as it develops vortexes. Engineers refer to this as transition flow. As the stream progresses it becomes white with vorticity and the vortexes become violent. The water flow is then fully developed turbulent flow. In summary: The flow is; 1) laminar, smooth and clear. 2) in transition from smooth to turbulent semi-transparent, and 3) turbulent mixing in the non-transparent, opaque phase. Strong turbidity with minute air bubbles makes the water's appearance change from semi-opaque to white water. (Photo courtesy Nancy McGehee)

For fluid flow in a pipe, the pressure loss per unit distance for laminar flow is much lower than the pressure loss per unit length for turbulent flow. Laminar flow pressure loss is almost directly proportional to velocity of the fluid, but the pressure loss in turbulent flow is directly proportional to the square of velocity. Pressure loss is in effect the increase in entropy. The flow pressure losses in the two differing flow conditions (laminar vs turbulent) can be different by factors of from about ten (10) to as much as fifty (50) or more. The mathematical relationships (Mathematical equation comparisons in Appendix 2.) were developed by engineers using empirical methods, but have been repeatedly proven to be valid.

A demonstration of Lower flow resistance for laminar flow as compared to turbulent flow was demonstrated by NASA (then NACA) by perforating the surface of an aircraft wing and then drawing off the turbulent boundary layer through the perforations by use of a vacuum pump (ejector). The experiment was successful, demonstrating that much less power was required to propel the aircraft. However, in their experiment, the power required for the vacuum pumping offset the gain. But the principle was proven.

In engineering fluid flow studies, pressure losses are related to velocities in confined flow channels, such as ducts and pipes. Equations relate pressure drop to density change and also to velocity. In our case there are no boundaries, and yet the total density of total mass per unit volume (#'s of mass per cubic light year) does not change. Conventional pressure loss equations do not directly apply for the deceleration of mass in the Stage III flow, because pressure loss equations relate to flow in confined ducts or pipes. In the universe where the mass and space are decelerating from the deflagration front, the following conditions are found:

1. There are no boundaries. The only place where the "fluid" can expand laterally is occupied by other 'fluid'. The only boundary-like region is the contiguous flow of previous time, which is also contiguous with more previous flow. The shear region is a velocity gradient that is perpetual.

2. The overall density of the fluid is constant. The total mass per unit volume of each large region of the universe remains constant. Only grain/cell sizes increase. As objects coalesce, the compound objects' gravities combine, but the overall separation distance between this and other compound objects becomes larger, which decreases the force from gravitational attractions.

3. The system is adiabatic which simply states that there is no energy added or removed from the system as the flow progresses.

4. Pressure loss in fluid flow reduces velocity without changes in fluid temperature.

5. The velocity (and linear momentum) decreases are system entropy increases.

Fluid flow parameters (mass flow, volumetric flow, velocity, etc.) are usually calculated by measuring the pressure reduction through an elbow, orifice, or other pressure loss device. In the flow of mass from the deflagration front, there not only are no boundaries, but there is no heat taken from or added to the system. Fluid flow equations simply state that pressure loss (entropy gain) is proportional to the square of the average velocity over the distance of the pressure loss. Graphically, such an equation yields a parabolic curve, as shown in the following charts. The unknown is how fast does the velocity convert to entropy? This is not known from analysis today, but in the future can be determined by analysis of data from red-shift and mass distribution surveys of the universe. In the meantime, "estimated" deceleration rates are graphically presented, which illustrates the process.

Pressure drop compared to velocity in turbulent fluid flow is a square function and therefore when plotted on a graph, it describes a parabolic shaped curve. For the preceding laminar flow the pressure loss per unit length of flow is almost a straight line relationship. In laminar flow, the particles are traveling parallel to each other. In turbulent flow, the particles within the fluid are no longer traveling parallel to the flow of the fluid. The deceleration curves are thereby steeper for turbulent (mixing) flow. For free jet flow in space, like the exhaust plume from a jet engine or rocket, flow starts as turbulent and turbulent mixing increases as flow distance increases, and the jet velocity continues to decrease, with the increasing entropy. This is the process which is occurring in the generation of the Stage III universe, but there are no known ways to measure the dynamic pressure losses. In addition to the described fluid flow, especially in the laminar–to–turbulent transition zone (Stage II C), there are significant high-velocity-enhanced-gravity fusing and coalescing of particles, which forms the first isotopes and nuclides, larger and heavier than Hydrogen (H_1) nuclei.

<u>*Stage II A*</u> *is the transition of matter from positrons and electrons to photons, which is from a mass state to an energy state; can be considered analogous to sublimation phase change from solid state, e.g., dry ice or water ice, to gaseous state, e.g., carbon dioxide gas or water vapor.*

<u>*Stage II B*</u> *is the universe's laminar fluid flow region. Fluid flow, both for liquids and gasses, starts moving as smooth and laminar (no turbulence): particles move along smooth layers in paths, with each layer gliding along adjacent layers.*

Stage II C *is the flow transition region, which is where the fluid violently transitions from steady laminar flow to increasingly turbulent flow. In fluid hydraulics this is referred to as the critical, which is where the engineering empirical term Reynolds Number, reaches the value of 2,000 (Reynolds number is a convenient value which is the algebraic product of density, velocity, and distance traveled, compared to the fluid's viscosity. The Reynolds number allows engineers to predict where the transition zone will occur). This transition zone is the flow region where the smooth laminar flow, through minute eddies and vortexing, transitions into highly turbulent flow. This phenomena can be observed in a creek or river where the flow is smooth and clear, and then over a short distance, the flow suddenly becomes turbulent. Shortly thereafter, when the Reynold's Number reaches a value of 4,000, the flow suddenly turns highly turbulent. For a water stream, the flow is often so turbulent it is referred to as white water. The transition region is quite short compared to the total flow.*

Note: It is suggested that the reader frequently reflect back on the pick-up truck load of potatoes from time to time to keep your mind's eye view correctly oriented relative to the proper red-shift interpretation. I do. Fluids are usually considered as volumes of large quantities of particles, whether a mix of smoke particles and gas molecules, or in liquid as numerous closely packed but free moving molecules. However, the smaller quantity of potatoes flowing out of the back of a truck is also a fluid flow. The potato flow out of the truck is a lumpy fluid, but the flow phenomena is similar, if not the same. In addition, it is also important to remember, that the potato truck example is a one-dimensional event. The New Universe Theory Concept 'reduction mechanisms' flows occur, and are continuing to occur outward, in all three dimensions, and in all directions simultaneously, as well as continuously.

The whole New Universe Theory Concept is a complex multi-stage fluid flow process. Conventional and well understood Fluid flow phenomena provide a framework upon which a computer model of the real dynamic universe can be developed. The likes of which we have never before envisioned and this is why fluid flow phenomena is introduced in this NUT explanation. A consideration about fluids to remember is that fluids can be in any one or any combination of three states; Solid, Liquid, and Gas, and these are each referred to as phases of state. In complex systems some fluid flows are multi-phase flows. They are a result of pressure and temperature changes, some due to velocity changes or energy inputs or extractions. All fluid flows are characterized by viscosity. Viscosity is the result of mutual attractions of the smallest of the fluid particles, the molecules.

Outline of the New Universe Theory Concept multi-phase, multi-stage flow process:

a. The Positronium crystalline solid primordial matter, when contacted by the energy laden deflagration front, by annihilations flashes positronium mass into photons, like the sublimation flashing of dry ice (frozen carbon dioxide) into a gaseous vapor.

b. Photons, like gaseous matter condensing into droplets, coalesce and condense into particles, most of which ends up as protons and neutrons. Like the coalescence of bubbles in a boiling liquid, some protons and neutrons combine and fuse into particles of larger isotopes and molecules. Of course by a completely different process, but similar in principle. The coalescence between particles is motivated by velocity-enhanced gravity which provides the force/pressure for proton to neutron, and neutron to neutron, fusions.

c. Fluid flow goes into what is called a transition zone after traveling a distance and the (continuous) inter-layer viscous shear results in velocity profiles such that small cells of the flow orbit in eddies and flow (relatively) in the reverse direction from the main fluid body. These eddies are the turbulence (vortexes) which propagate through the fluid. This correlates with Stage II C of the New Universe Theory. The viscosity (fluid cohesion) in the Stage II universe "fluid" flow is due to the velocity enhanced gravitational attractions between all of the mass particles.

d. As turbulent flow progresses the vortexing and combining continues. Pressure drop occurs more rapidly than in the previous laminar flow zone, because particles-orbiting-particles periodically come closer to other nearby particles for more influential gravitational interactions. For fluid flow in a pipe, the pressure drop is faster than in the New Universe Theory Concept Stage III, because in space there are no side boundaries to start the initial velocity profile changes. However, adjacent are more particles and more flow for mixing, and since the older Stage II fluid is traveling slower than the newer, there is a continuous shear stress. Velocity enhanced gravity reduces to levels inadequate to cause fusions, and more normal gravity only causes clumping into Stage III objects (stars, etc.) as velocity enhanced gravity decreases towards normal gravity. Shear stress continuously causes more vortexing resulting in more linear velocity reductions. This is the case of the turbulent universe's flow components of rotating galaxies, orbiting galactic clusters, and revolving structures of ever increasing sizes.

The following figure illustrates the flow lines during deceleration. In this graph Hubble number lines are linearly extrapolated and superimposed as illustrated in the preceding chapter in Figure 7.6 was an over simplification and the result is an under statement for the size of the universe. The parabolic (2nd order) decelerations of Stage III objects of the universe correlates with all other fluid flow phenomena. The initial flow velocity reduction of (Stage IIB) from the 100% c, by 3 to 4 % c from the annihilation zone (Stage IIA) to about 97 or 96% of the speed of light is mostly laminar flow, and has ~50 to 500 times less dynamic pressure loss for unit distance as compared to the turbulent flow region, the continuation of flow into the universe (Stage III). The first three or four % of radial velocity reduction distance is the laminar Stage IIB flow zone, but adds a significant distance to the continuously growing size of the universe. These considerations are illustrated in the next two graphs, (repeat of figure 7.6. with additional caption and a typical flow line through the transition flow zone, as illustrated in Figure 8.4). This extends our estimates for the size of the universe by adding ~2.5 billion light years to the radius (5 billion light years to the diameter), which is determined and explained in the next chapter.

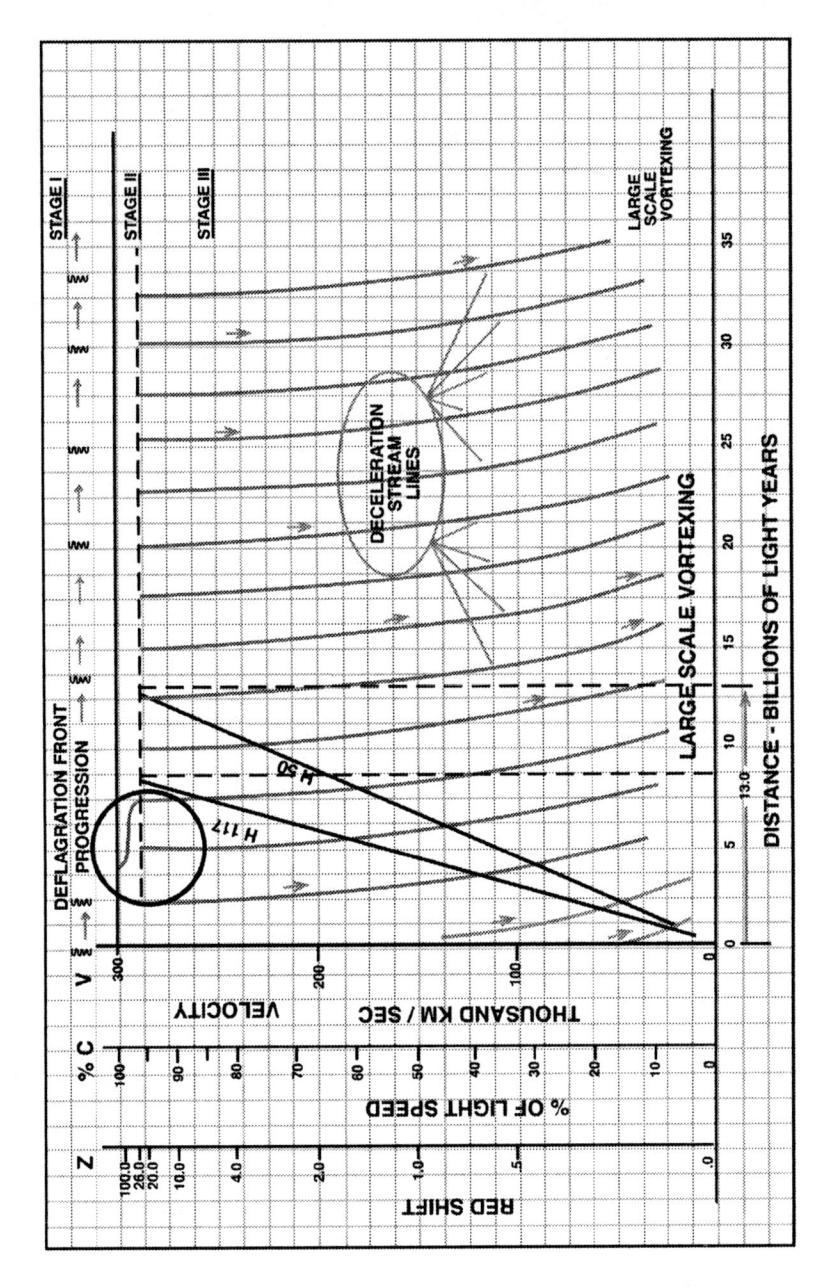

Figure 8.3. Universe Generation via Matter Flow. Stream lines are drawn to assist the visualization of the decelerating flow from annihilation, through the deflagration front. The first process is laminar, the second is where the transition from laminar to turbulence occurs, and the third is where the turbulent mixing clumping, coalescing, and orbiting

rotations occur in the observable universe. The flow is away from the universe, and the universe size continuously grows to encompass the added-processed matter. Deceleration rates throughout the process are graphically determined and found to be extremely slow. It ranges from one meter per hour in 130,000,000 kilometers, up to the deceleration rates of one kilometer per hour in 36,000,000 kilometers. (Compare deceleration of your automobile when approaching a stop light). The Hubble lines (H 117 and H 50) are velocity gradients looking upstream (two directions) across stream lines.

Mixing and decelerations occur very gradually in the initial high velocity laminar flow region of the New Universe Theory of matter transformation. During this highest velocity region there are large quantities of mass particle fusions due to very high enhanced gravity forces, to and by all particles. As particles combine into multiple hadron isotopes of nuclides, the flow becomes complex multi-phase flow. It is anticipated however, after extensive analyses and follow-on computer modeling it will be analytically possible to more precisely define the shape of flow lines for the laminar and transition flow zones. In the meantime the initial flow line curve shape is an estimate, based on conventional fluid flow phenomena. The Stage II best-guess flow curve is illustrated in the 'zoom in' view of the circled area from Figure 8.3, in Figure 8.4.

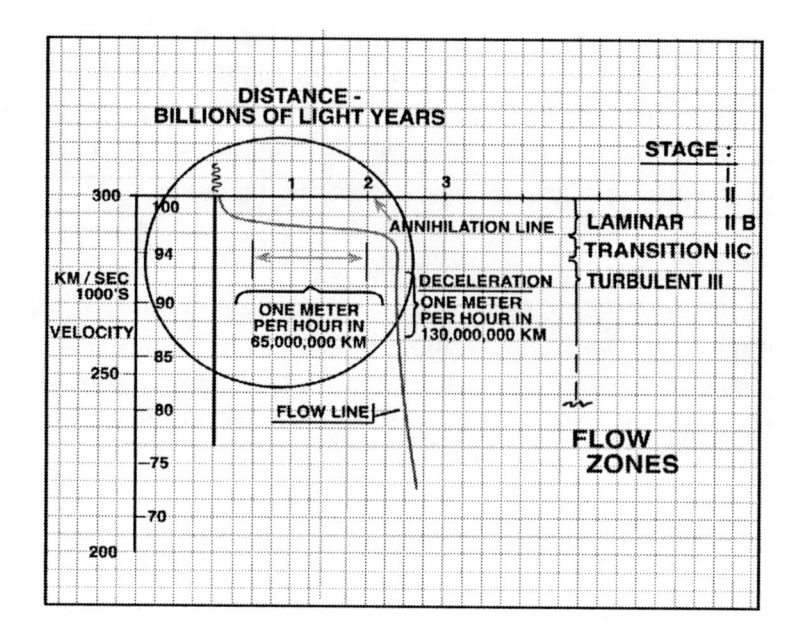

Figure 8.4. Laminar & Transition flow. This 'zoom-in' view of a flow line shows the (typical?) flow velocity profile of matter transitions from Stage I annihilations, through the generation mechanisms of Stage IIA, IIB, and IIC, and then with the continuing flow into Stage III. It is estimated Stage III starts at about 96% to 94% of light speed.

In Figure 8.3, the two parameter limiting Hubble number lines of H50 and H117 were superimposed as if they are straight lines (and therefore they are not valid as drawn. The Hubble Lines are revised from linear to reconciled curved lines and are presented later-on a chart in Chapter 10). If all these numbers and this tremendous size of the universe don't make you feel small and insignificant, review and re-look at the local universe picture from Figure 1.1, shown as a one billion light year sphere, with our galaxy located near the center in the Virgo clusters .... Now review the data graph of pressure drop (entropy increase) curves plotted over the Hubble number lines of Figure 8.3. The red-shift of objects observed along the Hubble lines have greater red shift with distance, but the objects (galaxies, quasars, etc.,) are decreasing in speed along the paths illustrated by the curved flow lines. Recognition of these processes is what initially prompted me into developing and revealing the NUT.

Pondering and studying the preceding fluid flow graphs reveals why the concept of the New Universe Theory results in an "Astounding Celestial Revelation"! Keep on reading, there are more exciting discoveries to come.

Chapter 9 ... Size & Age of the Universe 20th Century Version

This Chapter 9., describes and summarizes the age of the universe, based on current wisdom, which has been limited to 20th century knowledge. Those age components are valid, but only for age building blocks of the universe, and therefore those answers are only part of the age of the universe. *(e.g., the age of a nation cannot be determined by researching and knowing the age of one of its towns).* The following Chapter 10., reveals our universe's age based on the 21st century New Universe Theory. For those who are up-to-date with present universe age study and research, this Chapter 9., is a summary of currently understood knowledge of dating techniques, but it also relates that information to the NUT.

Measurement of anything in astronomy usually requires use of several different scales because of the large range of the numbers involved. Age, from time-zero (different from that based on the BB) till now, has taken a long-long time and requires several techniques to measure the large ranges of time and distances. The currently thought age of the universe is determined by summing the ages of the several phases and stages of development of the Milky Way galaxy. Evaluating each age dating component has required the specialty and expertise from several technical disciplines: Physics, Engineering, Chemistry, Astronomy, Geology, Cosmology, and others. All disciplines are significant participants and contributors to age study.

Radio-Active decay is one of the essential techniques. Carbon dating is a radio-active-decay technique, but is only accurate for dating ages of a few thousand years. For astronomy, this is not adequate, but it is one of the radioactive techniques of which most have at least a cursory awareness and it is a scientifically sound and it is very accurate. By describing this, all can better grasp how, in general, all the radioactive decay 'age dating' techniques work. Most are aware of the technique, but since it is not generally understood, here is a quick overview. All carbon isotopes have 6 protons in its nucleus. Carbon 12 is the most common and most stable isotope of carbon and its nucleus includes 6 neutrons, but the nucleus of carbon 14 isotope has 8 neutrons, for a total of 14 hadrons. Carbon 14 has proven very useful in dating biological things. For the last several million years, (possibly as many as 250), carbon 14 has been in a steady state concentration in the earth's atmosphere. Carbon 14 is manufactured in the atmosphere by solar radiation effects on the more stable carbon 12. Plant life on earth depends on and consumes carbon dioxide from the atmosphere and plants through photosynthesis, convert the carbon into plant growth. The primary element of plant life is carbon (and therefore animal life), and the only source of carbon for plant growth comes from carbon-dioxide in the atmosphere. Animal and vegetable life are based on carbon chemical processes. The ratio of carbon 12 and carbon 14 in the atmosphere is balanced in a steady state process, which implies the production rate and the decay rate are equal.

(Isotopes of many elements gradually decay back towards the smaller basic isotope components from which they were

originally fused in the processes by the stars. The solar system is a product, accreted from the supernova debris from much earlier stars, which originally consisted primarily of hydrogen. The larger stars were so massive their own gravitation caused very high pressures and temperatures in their large thermonuclear core and they fused (burned) their hydrogen reserves very rapidly; The products of combustion were heavier isotopes. The larger the stars were, the faster their fuel was consumed. The first isotopes were fused into yet heavier isotopes; each step converted some of the mass into energy which was radiated as light into the surrounding universe. When the mass reduced so rapidly in large fast burning stars, the corresponding rapid gravitational pressure reduction resulted in a supernova explosion which produced and disbursed into dust and gas clouds into the surrounding space, and these debris are known as nebula. The explosions were so violent that all the heavier than iron elements are actually produced by the high pressure waves in the explosion. In time, the clouds contracted, and in a few or less billion years, coalescence produced secondary stars, some with planetary solar systems. Our star and planetary system is a second or third generation product. Decay of isotopes is a random statistical process and with so many trillions of particles in decay, the average rates of decay are precisely predictable.)

Any given quantity of carbon 14 decays at a rate which results in one half being converted back into carbon 12 every 5,715 years. This time is known as the "half life" of carbon 14. When a biological being or substance dies, its consumption of carbon stops. However, the continuing decay of carbon 14 results in a reduction of the concentration ratio of carbon 14 and carbon 12 in any dead test specimen. Measurement of the ratio of quantities of carbon isotopes allows calculation of percentages of the isotopes, and thus the length of time since the subject's consumption of carbon stopped. This method is a very accurate dating technique for periods up to about a half-dozen Carbon 14 half lives, (about 35,000 years).

Many isotopes (atoms of all elements) decay over time towards more stable and smaller/lighter mass isotopes. Some isotopes are so stable that it may take '<u>almost</u>' forever for them to decay. For measuring long ages, elements and isotopes with half-lives in the range of the time period of interest are used for dating. Heavy elements like uranium 238 and lead 206 (one of its products of decay), ratios are useful in dating earth rocks and meteorites. Other isotopes are used for dating rocks that do not contain uranium.

Earth's oldest rocks have been found to have an age of **3.9** billion years. Some meteorites have been found which are as much as **4.56** billion years old. This implies that the earth rocks solidified 3.9 billion years ago and solar system formed 4.56 billion years ago. The supernova that manufactured the fragments for **our Sun**, a second or third generation star and **our solar system** occurred 4.56 billion years ago. Coalescence, vortexing, and clumping (precipitation) requiring the time difference of about 660 million years.

With spectral analyses of gases seen in the Milky Way, ratios of some isotopes can be determined. Rhenium 187 decays into Osmium 187 with a half life of 40 billion years. According to a paper by Edward L Wright, 11 June 2002, 15% of the original Re has decayed, which calculates to an age between **8 and 11** billion years (for the data source region of the Milky Way Galaxy). (Wright describes this as the age of the universe, but we now know his age calculation is only a small portion of the universe; actually, only the age of the observed gas), he would have been accurate if he stated "the age of observed gas found in the region of the Milky Way galaxy."

Dating of other, more distant regions at the periphery of the Milky Way galaxy requires another measurement technique. For many years physicists and astronomers have been studying the physical processes taking place in various sizes and types of stars. The life cycles of some are now well understood. Hydrogen to helium fusion (burning) is the energy source for all stars at least in their early life. Two physicists, Danish Ejnar **Hertzsprung** and American Henry Norris **R**ussell, plotted stars on a graph of luminosity verses temperature and found that most stars fall on a narrow line grouping, which they named the main sequence stars. The graph is known as the **H-R Diagram**. Within a class of stars, the luminosity decreases with increasing age. Various astronomers have found the age of globular clusters to be **within a range of 12.9 to 16.3** billion years old. This is somewhere between 2 to 6 billion years older than our galaxy stars, and provided a good "ball park" estimate, but these ages are not as accurate and definitive as we would like.

A recently developed dating technique appears to be quite accurate. It involves a type of star known as a 'white dwarf'. White dwarfs are the remains of specific type an size stars, similar to our sun. They have a life of about 8 to 10 billion years, and they then evolve into a degenerate white hot core. After that, their cooling rate is a simple heat transfer by radiation calculation. The measured radiation (light luminosity) from one of these stars reveals its age since it became degenerate. (Our sun is about 4 billion years of age, therefore we have another 4 billion years before we need to worry about its next stage of evolution.)

Astrophysicists have studied the evolution and aging processes of stars of many types and sizes. They have found that stars the size of our sun, and slightly larger, continue their fusion processes for about 8 to 10 billion years, until they have burned most of their hydrogen fuel in their core. Hydrogen outside the core continues to undergo reactions and expand. By simplified explanation, the star goes through two or more red giant stages. The core cools and nuclear reactions intermittently cease and restart due to lowering of pressure and temperature. Then, through 'planetary nebula' stages, the star sheds more mass through multiple pulsations of on-off-on fusion cycles, until all fusion stops, leaving only a white hot degenerate core, which is a 'white-dwarf'. It is a sphere smaller than planet earth, but brighter than our sun. White dwarf stars can cool only through radiation, which takes several billion years. By measuring the current luminosity (radiation level), past cooling time can be calculated with reasonable accuracy. Stars that start their life at sizes significantly larger than our sun also go through red-giant phases. Those stars go through a supernova stage which usually results in a more dense core than that of a white dwarf. They result in either a neutron star or a black hole. According to a press release on August 28, 1995, a team led by Harvey Richer from University of British Columbia, using the Hubble telescope took some 5 hour long exposures of the faintest, coolest white dwarfs in the M-4 Globular Cluster (closest globular to earth, only 7000 light years). The long exposure times assured obtaining their data from the faintest and therefore the oldest observable stars. They found the results that implies the <u>M-4 cluster's age to be **13 billion years**</u>. (Question; How do they know there aren't older white dwarfs that are too faint to see?) Their data implies the globular clusters are about 2 to 4+ billion years older than the Milky Way galaxy.

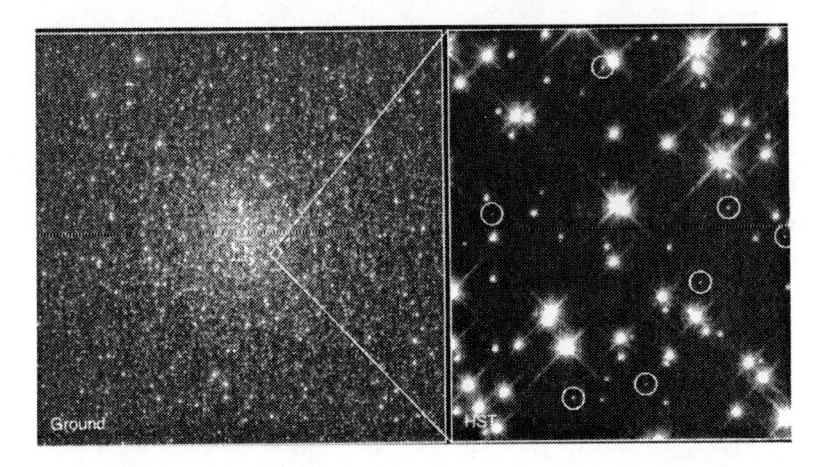

Figure 9.1. White Dwarf Stars in the M-4 Globular Cluster.
Age of our vicinity of the universe is shown by Harvey Richer and his history making team. They identified some of the oldest observable (they thought to be original) stars in the M-4 Globular Cluster and then calculated their age by measuring their brightness (dimness) using the Hubble space telescope.

Members of Harvey Richer's history making team were Gregory Fahlman, Rodrigo Ibata, and Georgi Mandushev (University of British Columbia); Roger Bell (University of Maryland); Michael Bolte (University of California-Santa Cruz); William Harris (McMaster University); James Hesser (National Research Council of Canada); and Don Vandenberg (University of Victoria).

Bobby McGehee

The oldest known white dwarfs in the disk of the Milky Way galaxy were found by Oswalt, Smith, Wood, and Hintzen; reported in Nature in 1996. These faint Milky Way white dwarfs were found to range between 8.7 and 11.6 billion years of age. Which shows the region of the Milky Way white dwarfs to be slightly older than previous galaxy dating techniques, but younger than white dwarfs in our satellite globular cluster M-4.

Those results are considered consistent with the earlier dating of our galaxy. Also, according to the "Standard model" of the BB, and as described by Steven Hawking, who theorizes the time required for the formation of the first star is only about one billion years.

The ages of various components of the local universe are as determined by physicists, astronomers, chemists, and geologists, and were believed to culminate into the age of the universe. That was under the BBB assumption that all of the universe was formed at one time, which the New Universe Theory now postulates as not valid. For a starting place, the ages determined from the described techniques are as shown in Figure 9.2 table.

NEW UNIVERSE THEORY WITH THE LAWS OF PHYSICS

Proven Ages in the Vicinity of the Milky Way (MW) Galaxy		
Component of Universe	**Time from Component Origin**	**Dating Technique**
Earth	3.9 Billion years	Rock Radioactive Decay
Solar System	4.3 Billion Years	Meteorite Radioactive Decay
Milky Way Galaxy	8.7 to 11.6 Billion Years	MW Gases spectral Analyses
Globular Cluster M4	13.0 Billion Years	White Dwarf Stars

Figure 9.2. Table. Proven Age Components. This table identifies the milestones that shows our earth as the youngest in this list of components of the universe; and Globular Cluster M-4 as the oldest. When the BBB was still accepted as valid, the oldest component age in the table, and including Steven Hawking's 1.0 Billion years (time for the first Population II star formation) was assumed as the age of the universe. That age was 14.0 Billion years.

Prior to revealing the New Universe Theory, the best available estimate for the age of the universe was equal to the globular cluster age plus the estimated time for formation of the first star, estimated at one+ billion years, based upon the mathematical and computer model of the hypothesized BB. The then concluded total age was 14 billion years. According to the assumptions, prior to the debut of the NUT, it is said the radius of the universe is less than an estimated 8 to 12 billion light years. It was assumed the universe accelerated from a single dimension-less point to 8 or 12 billion light years in 14 billion years!

From this time forward, the New Universe Theory provides more, exciting new adventure, making available new understandings. After reviewing all of the above previous **big** numbers, it is time for us to get the rest of the story! Take a deep breath, and proceed through the next chapter, which provides an enlightening revelation.

Chapter 10 ... Size & Age of the Universe
21st Century Version

As Paul Harvey might say;
"Here is the Rest of The Story!"...
Age of the Total Universe; How old ?: How large ?

The Universe's growth is on-going. The NUT explains how primordial matter and space are continuously being processed, adding more universe in the same proportions of space and matter as currently exists. As all physicists and readers of this New Universe Theory know, matter includes mass and energy. The deflagration front is continuing to advance through more primordial matter, simultaneously processing more mass, energy, and space into the universe. By re-examination of astronomy knowledge, and in the light of the NUT, we discover how gigantic the universe really is. In the preceding Chapter 9, logic and current knowledge reveals the basis for knowing the age of <u>our region</u> of the universe. Universe size and age analyses, under the hypothesis of the BBB, that all matter appeared simultaneously and instantaneously. It was further assumed that the age of our region of the universe is typical. Those assumptions have been shown by the NUT to be erroneous. The age 'building blocks', as presented in Figure 9.2 Table, are accepted as facts, proven by the Physics, and the Astronomy communities. Since the well conceived and executed study in 1998 by Harvey Richer's history making team, it is recognized that most astronomers and cosmologists have concluded that the 13 billion year age of the M-4 white dwarfs, as the time since the universe and the Milky Way Galaxy region of the universe came into existence. My objective now is to show that the <u>whole</u> universe is much older than our Milky Way Galaxy. Most of it is younger, yet some of it is older. Also, the NUT includes an estimate for primordial matter processing time.

Time and distance Factors:

(A) <u>The length of time and distance from the beginning of the deflagration front until the time when and where the Milky Way region of the universe first came into existence</u>. That is the time between the initial start of the universe generating deflagration Front, and the time when the Front passed through this region. (Like a half loaded truck spilling potatoes along the way). The time of travel between initiation of deflagration and when it passed through our region, has been graphically determined from Figure 7.6, and is illustrated in Figure 7.7, graph in the Chapter titled Deceleration and Entropy. The distance transcended by the Front during that time increment was initially and tentatively determined to be the <u>distance</u> of 3.5 Billion Light Years, and since the deflagration front travels at the speed of light, that distance number corresponds to the <u>time</u> increment of <u>3.5 Billion years</u>.

The distance from the starting place of the universe to where our region started is based on the Figure 8.3 nomograph distance from the point between end-to-end drawing of the smallest and largest Hubble number distance base lines, and the midpoint of the sum total of these two Hubble number base lines. Reconciling is accomplished by curving the H117 Hubble line (Figure 8.3) to the distance from our location to the nearest point on the universe's periphery, which corresponds to the Front travel distance during the aging of the M-4 white dwarfs (H50 is curved proportionately as shown in Figure 10.1). Now, the midpoint of the sum of the distance for reconciled maximum and minimum Hubble lines is where the universe started, since the Front travels the same speed in all directions. The point between the end to end max and min Hubble base lines is our location. Of course, this is from where the Hubble numbers were measured. The validity of the Pre-Milky Way (PMW) value depends upon the true, yet to be determined (confirmed, and verified) maximum and minimum reconciled H lines. Hopefully this will be established in the near future from reduction of data from the several recent red-shift surveys of the universe, when compared to known and directly applicable distance candles. Distance candles for red-shift calibration beyond one third the speed of light will have to rely on direction unique distance candles such as brightness of specific types of galaxies (spiral type ?), in the vicinity of the red-shift measurements. Other distance candle methods at these distances are simply too dim. New technology telescopes will contribute, and hopefully be adequate. Until these tasks are completed, the best currently available reconciled nomograph value of **3.8 BLY**, will be used for the PMW time and distance.

(B) The processing time and distance for the reduction mechanisms to progress from the transition of the deflagration front until proto-matter is transmuted and mass is precipitated, coalesced, and ignited as the first observable objects. This time increment and distance is estimated and illustrated on the Figure 8.4 graph presented in the Chapter on Fluid Flow. Travel distance and time occurs while the new mass coalesces and flows through the front thickness of

<u>2.0 + Billion Light years</u> over the corresponding and slightly longer time increment of <u>~2.5 + Billion Years</u>.

This time increment has been graphically estimated but should and can be re-estimated by Fluid Flow knowledgeable Particle Physicists. It is expected that several such estimates will be forthcoming, many of which hopefully will be better than mine. It is unlikely that there will be a single value that will receive consensus throughout the Cosmological community until a complete and generally accepted computer model is developed that can define the yet to be accepted "Standard NUT model". It is also expected this will require a few years, and until this is done, my estimated value of **<u>2.5 Billion</u>** can continue to be used.

(C) <u>Another increment of time takes place; from when the first Sol size star ignited, until it became a white dwarf.</u> That is the time until it consumed a considerable portion of its nuclear fuel, to the extent that it collapsed, expanded into a red giant, and then through convulsions, ejected mass into planetary nebula, and ultimately left its remnant ..., a 'White Dwarf'.

Astrophysicists study and analyze the physics processes in stars. In particular the precursors of white dwarf size stars. They have determined that the time for Sol (our sun) size stars, the nuclear fusion phase is about 10 billion years before sufficient fuel consumption leads to star inflation into a red giant. The star then exists as a red giant for about .4 billion years and then the star remains, after convulsions and mass ejections into planetary nebula, as a white dwarf. In addition to the preceding summarized 10.0 billion year normal star nuclear fusion life, and .4 billion year red-giant life, the white dwarf requires an additional 13 billion years to cool to the 'dimness' level Harvey Richer of University of British Columbia and team observed. (This team made a significant contribution towards knowledge of one of the time increments for the age of the MWG). Richer called this the age of the universe, but with due respect, this is the age of only the stars observed. We do not know how long it has been since the first star cooled to the level of dimness that we cannot see it at all? The age of the immediate region of the M4 globular cluster must also include the aging time for the precursor Sol size star.

The life span from the precursor star ignition to the white dwarf fading from view is, in addition to Richer's 13 billion years, the total of all the preceding time increments, which adds up to an <u>age</u> increment 23 billion years. The calculations for the life span of these stars is based on technically sound analyses, which are accepted throughout the Physics and Astronomy community. This total white dwarf process is a significant portion of the <u>time</u> for the development of the MWG region of the Universe.

The <u>size</u> estimate for the universe is directly based on the product of the speed of the deflagration front (light) multiplied by the time when the deflagration front passed from its place of origin through primordial matter until the present. This is significantly different from the MWG distance from deflagration start site. As we know the mass objects slow in linear motion as their momentums are transferred into angular momentums (spinning, revolving, orbiting; of stars, star systems, galaxies, galactic clusters, super-clusters, and 'structures'). Graphical analyses as illustrated in Figure 10.1., shows that our MWG location is still less than about 4.0 billion light years from the origin site of the Deflagration front, and our linear velocity has continued to decrease towards zero. (I think our linear speed can be calculated from analyses of yet to be acquired direction calibrated Hubble numbers). The 'deceleration stream lines' in Figure 10.1. are simply my estimate and further thought suggests that they might even be steeper than shown in the Figure. I may have even drawn the lines curved in the wrong direction. These lines should probably be parabolic, with the lower end of the curve becoming indeterminate as the objects linear velocity approaches zero speed.

(D) <u>Significant is the distance the deflagration front (NUT shell) traveled to the current periphery of the universe, during cooling of the White Dwarfs.</u> It is (has been) assumed that <u>all</u> Milky Way globular clusters and their stars are about the same age, ... <u>not yet demonstrated, but temporarily accepted as OK, for now.</u> Since the NUT shell grows outward at the speed of light, the radial growth of the universe's size during the aging of the observed white dwarfs is also <u>13 Billion Light Years</u>.

The graphically defined distance from us to the universe's periphery at the deflagration front (presented in the Chapters on Entropy and the Fluid Flow) must now be revised to reconcile with the white dwarf defined minimum age of our region (Figure 10.1). Hubble number lines must be curved to correspond. It is now known H numbers are not uniform with distance, even in one direction. Hubble number lines are linear only for about the lower one third of their projected length; Beyond the distance associated with the velocity of 100,000 Kilometers per second, the line requires reconciliation. Adjusting the graphical illustration of H lines 117 and 50, as curved Hubble lines, reconciles the graph.

*********** **Special Note from the Author.** ***********

Our Universe's Size is the distance the deflagration front progressed during the sum of the above described Milky Way Galaxy known aging increments : (A)3.8 + (B)2.5 +(C)10.4 + (D)13 = 29.7 Billion Light Years. This radius is erroneously shown in Figure 10.3 as 19.7 BLY.

NEW UNIVERSE THEORY WITH THE LAWS OF PHYSICS

In finalizing the NUT for publication, I became aware of the error which is explained in the following paragraph:
I don't blame anyone but myself for overlooking the time increment of (C) is 10.4, not just .4 as used for my original calculations for the size of the universe. As a result, Figure 10.3 illustrates the diameter of the universe including the deflagration wave as 39.4 billion light years, when it actually is 59.4 billion light years. I believe Professor Durbin would still give me a grade of A- ! Revising the Figure 10.3 graph and a few related errors would possibly cause several months delay. So in the interest of getting this new theory 'out there' for consideration and kibitzing by others, I decided to publish the NUT with the 10 billion light year error. I accept full responsibility for that error.

**************End of Author's Special Note*************

A Red-shift scale is included on the Hubble reconciliation graphical chart along with the scales of velocity in both % of the speed of light, and as speed in kilometers per second. (Figure 10.1) For clarification about red-shift numbers, a description is provided in Appendix 1, but a brief note is appropriate. The Doppler effect is calculated by measuring the change in wave length of the observed light. The change is divided by the reference wave length and this provides the Doppler value (Z). Red-shift (Z') is then equal to (Z+1). For example; if the wavelength is stretched by a factor of 2.0, the change divided by the original, Z calculates as 1.0. (If you are interested, in more explanation refer to Appendix 3.0).

Since we are not located at the point of origin of the deflagration front that is generating our universe. Astronomers must now consider the angle between the object's recession and observation. The Doppler equations must now include the trigonometric function for the angle for the calculation of velocities. For simplification and to

avoid confusion, the graph (Z') does not include Doppler shift correction for the angle (Figure 10.2). The angle is also different with changes of distance. Also, another factor that must be considered is gravitational-red-shift, which in Stage II and early Stage III is a factor that occurs as a result of velocity enhanced gravity, as calculated by Einstein's Relativity Law. The complexity of calculations for the considerations mentioned in this paragraph are outside the scope of this work. Some additional discussion is presented in more detail in the Appendix.

Relief for astronomers is that the red shift still, as before, provides reasonably good velocity difference measurements between two objects, one the light source and the other the receiver. H numbers vary in each and all directions, and the variations are more pronounced beyond about 100,000 Km / sec, (one third the speed of light). The calculations of red-shifts to correlate with distances will require a calibration of the **"H--Line"** in each and every direction. Of course, for correct and accurate distance calculations it is essential that the correct H number be used for that direction. Past astronomical data obtained within the 100,000 Km / sec velocity is not expected to require significant corrections, except for direction considerations.

The curved portion of Hubble Number Reconciled Lines between the astronomer measured lower 1/3 H number and the M-4 known white dwarf age end point are estimated and free-hand drawn by the author on the chart of Figure 10.1. (This chart does not show direction, only distance and velocity.) Reconciliation is based on the age of Milky-Way vicinity globular cluster white dwarfs. This chart also shows the flow <u>lines of deceleration</u> which are followed by all objects from their initial transformation from energy-matter to mass-matter. Observation of a series of objects indicates higher Red-Shift (velocity) for farther distance, just as first discovered by V. Slipher in about 1912. Red-Shift is a measure of an individual object's instantaneous velocity relative to us, and does not in any way indicate any celerity changes (acceleration or deceleration.) Actually, <u>all objects are decelerating</u> from the deflagration Front **(NUT Stage II reduction mechanism shells)**, where the objects started their existence at the speed of light.

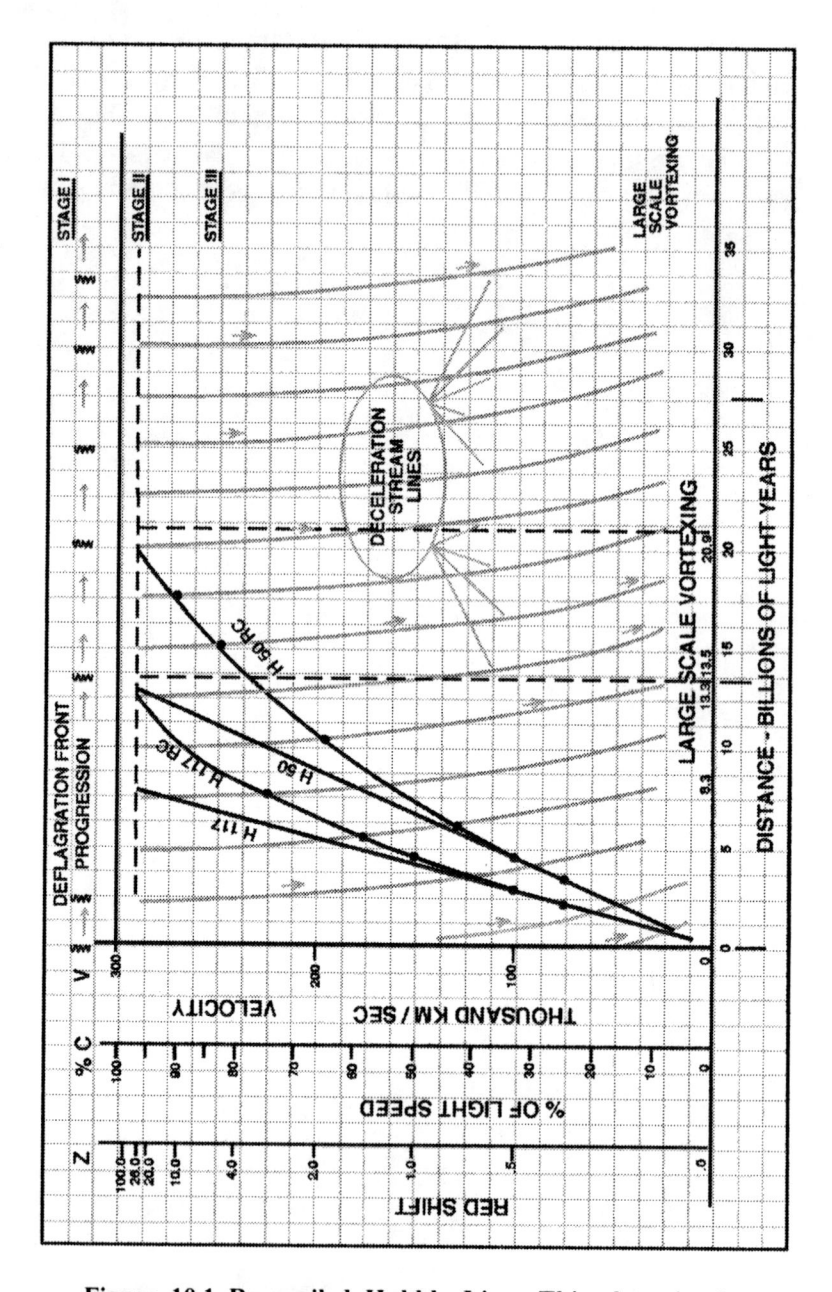

Figure 10.1 Reconciled Hubble Lines This chart is also a nomograph that allows astronomers to determine the distance to an observed object. On the left is a scale of Red-Shift which is calculated with the use of spectrographic data for the object's radiating light compared to a spectrograph

of identical chemical elements on earth. Alongside the Red Shift scale are scales for determining the observed object's Percent-of-Light-Speed and the Velocity in Kilometers per second. By moving across the chart from the measured Red-Shift to the intersection of the appropriate H-Number-RC line for the direction of observation, then directly below on the chart is the distance scale, the distance to the observed object is thereby known. (e.g., the farthest object ever observed is said to be at the red-shift of 26.0. The nomograph shows that object was approximately somewhere between 13.4 and 21.0 billion light years distance, depending upon the direction of observation, which determines the applicable H-Number-RC line.)

A series of 'line-of-sight' objects will indicate higher Red-Shift (velocity) for farther distance. Red-Shift measurement allows calculation of the instantaneous rate of separation between a light emitting object and a light receiver. There is no known way to extract 'rate of change' (acceleration or deceleration) of distance directly from red-shift data. Also, red-shift provides only relative velocity information, and does not provide any absolute velocity information (except for objects observed along one specific line that radiates from the universe origin and projects through the MWG). Red-shift data is only applicable for the line-of-sight velocity vector. If all three direction components, and absolute velocity are known for one of the two (receiver and/or sender) from other reference information, only the line of sight direction and line of sight absolute velocity component could be calculated for the other using red-shift plus common trigonometric functions. At this time, how to determine the transverse vector component of either the receiver (MWG) or the sender is not defined. This is the main reason for the inconsistency for Hubble numbers between competent astronomers. Direction and angular position of astronomical light source objects relative to us change so slowly over these great distances, we cannot detect, let alone measure these changes with time. However, if we know our distance and travel direction from the Front Start Site, velocity data for other objects with travel along our traverse line (dashed line in Figure 10.2) will not require any corrections.

Calibration of the reconciled Hubble lines can begin with the newest distance candle; **Supernova 1997ff**. The highest red shift ever measured for a supernova Type 1a, was measured at **red shift of 1.7.** (Data from ref 44) This **corresponds to a velocity of about 55% of the speed of light**. If I knew the direction coordinates, I could adjust the reconciled lines on **Figure 10.1** to the corresponding distance calculation based on visual brightness of Supernova 1997ff. This calibration and others will be the first step towards calibrating Figure

10.1 into a nomograph for astronomers, for the first time, to have distance calculations at such high red shifts. They will be technically sound, being based on the New Universe Theory.

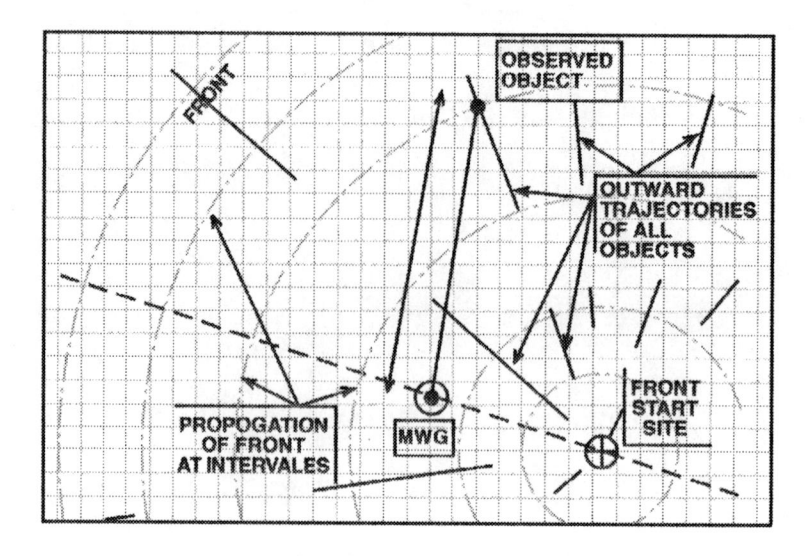

Figure 10.2 Velocity and Relative Velocity The rate of change of the vector (arrow) is the separation velocity as calculated from Red-Shift of the light coming from an observed object, and as measured in the Milky Way Galaxy (MWG). Actual as object velocity occurs during its deceleration as it progresses from, yet towards, the deflagration front in its outward trajectoryalong a radial line from the initiation site in the front. The front start site is a huge void as everything is decelerating away from it and from each other! Objects do not decelerateindividually, only as parts of mutually revolving and interacting systems, therefore decelerationrates could not be measured directly, even if we had the technique sensitive enough to do so.

Size and Age of the Universe.

The **astounding size and age of the universe** are illustrated in the following graphical drawing, (Figure 10.3) which reveals one half a cross section view. This section view drawing is through the spherical universe on a line that passes through the Milky Way of the Local Group Galactic Cluster as well as through the center of the universe. This view was derived from the chart of the **H**ubble **N**umbers **R**econciled **C**alibrated lines. One conclusion is that we are 3.8 Billion Light Years from the center of the universe, which indicates the vicinity where the Milky Way was generated by the deflagration front 3.8 billion years after the NUT Deflagration-universe-generating mechanisms started. The illustration of the Local Universe, (shown in the Chapter 1 Introduction), is drawn to scale in this graphical drawing of the Universe. The NUT **Stage II** Reduction Mechanisms shell is shown as 2.5 billion light years thick. Within a couple of years after the revealing of this NUT, a better estimate of this shell thickness is expected and anticipated after mathematical and computer models of the NUT are developed by elementary particle Physicists. (If Professor Alvin Pershing and Professor Durbin were around today,!......?) The diameter of the spherical universe that is within our range of view is shown as 39.4 Billion Light Years (BLY).

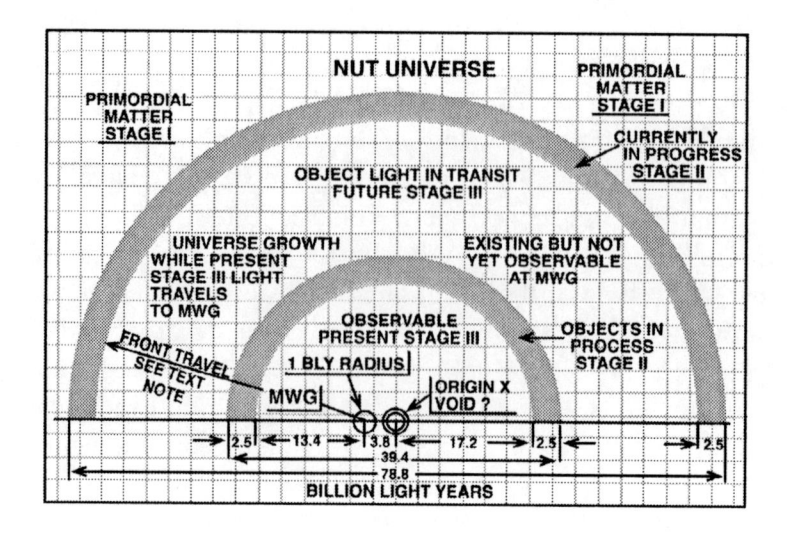

Figure 10.3. Size of the NUT Universe. Based on Reconciled Hubble Lines this view of the universe reveals the NUT size (and age) estimates for the universe. The inner semi-circle is the current 'known' universe size. The outer semi-circle shows where the periphery of the universe is today, assuming the primordial matter has not been depleted. While the light from inside the inner semi-circle was coming to us, the 2.5 BLY thick deflagration front will have advanced to the position of the outer semi-circle. The band width of the semicircle is the time and distance of the outward advancing deflagration front. (Steven Hawkins published a book titled "The Universe in a Nutshell". In his honor, we could refer to the deflagration front, a.k.a. the 'NUT-Shell.'

It was mentioned earlier that an object (Quasar) reportedly has been observed at a red-shift of ~26+. This corresponds to the very beginning of the Stage III universe, which is the beginning of star ignitions. Assuming that data is valid, and assuming that data was acquired and observed in the general direction of the nearest perimeter location to us, (in the direction of the Abell Clusters). The object's distance would calculate to be 13.3 BLY. If the object was measured in the opposite direction the distance is 20.9 BLY. (The observation was using light that left the object somewhere between 13.3 and 20.9 billion years ago). The progression of the deflagration front during the travel time to us by the object's light, and the object were both traveling (in opposite directions) at near the speed of light, the universe shell diameter would now be twice the observed object's distance plus the pre-MWG distance to the center of the Universe, plus the thickness of Stage II on both outer sides of the universe's sphere. These huge numbers are valid, assuming the NUT primordial matter reservoir Stage I, did not become depleted along the way. If primordial matter did become depleted occupants of the MWG might not know it for another 21 billion years!.

The Figure 10.3. illustration not only shows the universe that we can and are observing, but also shows the additional universe volume that has since been generated but is still too far for its light-in-transit to reach planet earth. However, only a small portion of that volume will be observed by earthlings, because our sun will deplete its fuel reserve in only ~ 4 billion years. Sadly, our sun will have died of old age several billions of years before light from most of such distant sources will arrive in what was 'our' region of the universe. Nevertheless, we humans can be proud to have provided the universe with animate intelligent consciousness for at least a small period of time during the universe's existence. Our universe is giving us the opportunity to see, and is saying to us, "Open your eyes and look at me". We are becoming reasonably intelligent beings, so if we try, we might be able to perpetuate the universe's consciousness via our vagabond descendants; intelligent robots.

NEW UNIVERSE THEORY WITH THE LAWS OF PHYSICS

Universe Age Before, In, and Beyond our MW Galaxy
From start of Universe

Front Travel distance	*Time Increment & (Total Age)*	*Activity During Time Increment*
2.5 BLY	(2.+) Billion Years (2.+)	Initial/First Photons to hadrons to nuclides
.001 BLY	+ (100) Million Years (2.001+)	Coalescence+Vortexes, Forming Stars, Globular Clusters, Galaxies
4.2 BLY	+ (4.2) Billion Years (6.201+)	Universe generation before MWG region existed
13.2 BLY	+ (13.2) Billion Years (19.401+)	Deflagration Front Progress, Nearest MWG Distance
13.2 to 20 BLY Direction dependent	+ (12 to 20) Billion Years (31.4+ to 41.5+)	Front travel since light left Farthest Observed Object (Red Shift=26+)

<u>Figure 10.4. Table. Age of Universe.</u> This table summarizes and presents pertinent data about the universes' size and age. These data are extracted from the cross-section of the universe as illustrated in Figure 10.3.

Age of Universe ... inside the NUT shell ... 71.4 Billion Years.

Amazing!

As often stated in this document, the deflagration front, at the periphery of the universe, continues to progress outwards in all directions at the speed of light. When we determine the age of the universe, it is a simple process to calculate the spherical size of the universe. (The total age of the universe multiplied by the speed of light, (one light year / year) which yields the same number in light years of distance.)

It is important to recognize that we are not at, or even close to the center of the universe. (Only our ancestral ingrained ego-centric psyche seems to desire that.) Our off-center position in the universe is illustrated in the graph (Figure 10.3); we appear to be in the general region of the central universe, the difference in distance to the periphery in opposite directions, is graphically shown to be about **3.8** billion light years. We are near enough to central that with our currently limited depth perception and observational resolution, things only appear similar in all directions.

For the curious amateur and professional astronomers, using the NUT concept and the Hubble number direction analysis, we can estimate the direction as well as the distance to the center of the universe. If the closest distance to the deflagration wave is towards the Abell clusters, then the farthest distance is in the opposite direction, which is also beyond the center from our Milky Way Galaxy perspective. The direction is about 3 hours, 22 minutes Right Ascension; and 35 degrees, 45 minutes Declination. In this direction we could find a huge void at the distance to the center, about 3.8 billion light years. From our simple graphical analysis, it can be estimated that the void is probably somewhere between one and 3 billion light years in diameter. Wherever the void is located, finding the huge center of the universe void will be a monumental observational discovery. It is entirely possible that the void will never be found as it may no longer exist.

We all will be anxiously awaiting similar graphic analyses of more refined and accurate data catalogued by directions and distance in three dimensions, and then modeled for computer viewing. Govert Schilling wrote an interesting article published in the February 2003 Sky & Telescope magazine. Govert addressed the need and desire for a 3D sky atlas of all of the sky. All new data, as well as the older data, needs to be cataloged with direction verses red-shift measured velocity and calculated distance to where the data was observed. In the meantime we have to continue to make some bold assumptions to conduct even this precursory analysis. It must be understood that all data measured from earth, or from the Milky Way galaxy were not taken from the center of the universe. Graphically plotting the Hubble numbers at various directions, and extrapolating them to the speed of light, allows illustration of how far we are preliminarily estimated to be from the beginning point of the universe, which is also the geometric center. Many of the recent and currently underway Sky Surveys mentioned by Schilling are listed in Chapter 6. The percent of the sky explored (not surveyed) by all of the above projects is estimated by the author at less than 10 % of the known universe's 47,700 billion-cubic-light-years of volume. (My estimate)

A computer model is needed that includes all of the Stage II A, B, and C phases of the reduction mechanism before the time can be reliably estimated. In the meantime, my estimate based on logic and graphical analysis will have to suffice. Stage II A (photon to elementary particle) time; the Stage II B (elementary particle to isotopes, and molecular clouds) time; and the Stage II C clumping and vortexing from clouds and dust to nuclear furnace illuminating stars). Understanding these "reduction mechanisms" helps to understand that the time for Stage II must be a very long time. My conservative estimate that is included in the total age, is based on the conclusions reached in the section on "Fluid Flow" (Chapter VII). That estimate was about ~ 2.5 billion years.

Compare the BB with the Laws of Physics compliant NUT: The BB is said to start with all mass at zero velocity, and zero size, and is progressing towards but can never achieve the speed of light; it is expanding to its death. The NUT starts out at the speed of light and grows towards an ever increasing size universe, and the fragmented and coalescing constituents progress towards zero expansion (linear) speed. However, The New Universe Theory concept continuously adds space and volume, and our Universe continues to grow with vim and vigor!.

How many theories, before the New Universe Theory, about the origin of the universe (and our world) have been invented that served the ego-centric psyche of we humans? Answer; ... All of them, except Hermann Bondi's, Fred Hoyle's, and Thomas Gold's continuous creation theory and that theory was rejected as it failed to answer some basic questions. All except that idea kept putting mankind at the center? Now, as Dr Phil McGraw often says, now we can get real!

With the data that is now available from the several sky survey projects, analyses and computer graphing can be done to calculate and show the Hubble number map of the universe. Other 'reference candle' data must also be correlated with the red-shift data to calibrate the Hubble numbers.

Using the Chandra X-Ray Telescope, Object # SDSS 1030 - 0524, (a Quasar), was observed by William Brandt of Pennsylvania State University and 15 colleagues, to have the red shift of 26+. (This calculates to ~96 % of light speed).

Chapter 11 ... Questions & Answers & More Questions

Answers are given to some of the questions brought to the forefront as a result of curiosity about this "Celestial Revelation" and "New Universe Theory". **More Questions** is the last phrase in the heading of this chapter as all of these answers and other new ideas should be questioned. Hopefully this work will stimulate more thought and therefore more questions. Several unanswered questions are identified by the author and more questions are expected from all that give serious consideration to this NUT "New Universe Theory".

Several of the following questions came from my family, (wife Nancy, son Bobby Jr, daughter Sue, son-in-law Jim, nephew Bill, and his wife Monnie,). Some of the answers to their questions are at least partially discussed directly or related to material in the book. Answers and offered explanations suggested by others may be submitted as appropriate. Welcomed are other questions and suggested answers, which are solicited on the web site: WWW.NewUniverseTheory.com.

Any answers that are suggested must be scientifically credible, therefore must be compatible with the Laws of Physics, to be considered. Additional answers and explanations that the author deems appropriate will be periodically posted on this web site.

Q. *How did this New Universe Theory come to mind?*

A. Logic, analysis, and dissatisfaction with the Laws of Physics violating BB idea. The proper interpretation of the red-shift; ... the objects farther away are moving faster than the objects that are closer, and that red shift is a measure of <u>speed,</u> <u>not acceleration</u>. Acceleration cannot happen without an accompanying force.

Q. *If all objects are decelerating, not accelerating, from where did they come?*

B. The production mechanism (Reduction Mechanisms) came from something before us and behind us, in space-time, but it is advancing away from us, in all directions, ever since we came into existence.

Q. *How do you know that the objects are not accelerating, like as claimed in the BB theory?*

A. There are no forces detected or conceivable that could provide continuing acceleration as required by Newtons Laws of Motion, and Newton's Laws have been proven to be valid. To circumvent the laws, some have said, even claimed to prove, that there is some mysterious "dark energy" causing the hypothesized acceleration.

Q *Newton's laws of motion require linear momentum to be removed from an object in motion for it to decelerate. If the objects are decelerating, where is the momentum energy going?*

A.   Entropy. Produced from transfer of linear momentum to angular momentum, rotational mixing, and spinning.

Q   *What is Entropy?*

A   Entropy is a term used in thermodynamics to express the loss of energy from a system and this energy is not recoverable. It is also a measure of disorder in a system. It is not heat (heat is internal energy and is known as enthalpy).

Q   *Since we now know there was no BB, what about the inflation idea?*

A   Inflation will go away, as it was just an imaginative and creative idea to try to explain some apparent aberrations of the BB concept. Throughout history, as new discoveries are made and knowledge is gained and proven, the old concepts (like inflation) become part of the study of our history.

Q   *What evidence is there that the primordial proto-matter is made up of positroniums?*

A   None. Positroniums are only one of the combinations of matter and anti-matter that could supply the quantities of energy necessary to fuel the deflagration front. However, the only unit combinations of matter and anti-matter that are known to exist are positroniums.

Q   *How can you be so sure that the proto-matter is a crystal lattice structure of hexahedron geometry?*

A   Whatever the geometry, it must provide uniform

separations in all directions, and the particles must be in fixed relative locations. This description fits the chemical definition of a hexahedron crystalline structure. The most stable and hardest of all substances are hexahedron crystalline materials (e.g., Diamonds, Carborundum).

Q *Since it is known that a deflagration front started once in one location, why couldn't there be more than one deflagration front? And therefore, more than one universe?*

A Yes, there could be, and my guess is that there probably are more.. If two deflagration fronts collided in the vastness of the pre-proto-matter expanse, they would eventually merge. The matter that is converted from energy to mass in the combined universes would be decelerating from different directions. There is some evidence that this may be the case, which would explain some apparent paradoxes of galaxies merging, each with vastly differing red shifts These are called 'discordant' red shifts!.

Q *From where did all of the proto-matter and proto-space come?*

A Answer unknown. Like a quote stated in my Physics text on "Mechanics: Statics and Dynamics"; The solution/answer is left to the student! Some theorist will surely someday come up with a laws of physics consistent suggestion. If anything could exist forever, in space or time, it could just as well be assumed to be primordial matter.

Q *When will all of the proto-matter and proto-space be consumed by the reaction mechanism processes, and what will be the consequences?*

A   A great question. Un-answerable at this time. A question for eternity? Our universe would stop adding mass, but inertia of the existing mass will continue. Since under this supposition there is no primordial matter left from which to produce new mass, there will gradually be less and less mixing and the density of the universe will disperse into the outer void. The distances between objects would continue to increase and the gravity force vectors continue to decrease. The answer to this question gets back to; does the inertia of the universe's components become low enough that they will be overcome by the universe's 'center of mass' gravitational attractions, at which time gravity would initiate and perpetuate a future contraction? I don't know, but this is a solvable mathematical question. (The gravitational attraction between objects is directly proportional to the product of their masses and indirectly to the product of their distances, so there should be no contraction for the outer portion of the universe. However, I am guessing the answer is;....... yes, a portion of the universe will eventually contract and collapse, and the outer portion of the universe will continue to disperse. The total process would be like a gargantuan supernova that has been expanding for 30 to 80 billion years, with an eventual, central black hole, but at the earliest this will only occur after another 30 to 80 billion years.

Q   *When, (and if) all of the "fuel" is consumed, will the big squeeze begin?*

A   The time between depletion of the universe's proto-matter, and the start and feasibility of contraction is mathematically determinable. Mathematical models could be developed that will show the time between proto-matter depletion and start of collapse. When and if all of the proto-matter is consumed and all mass is inside the universe, and after linear momentums are sufficiently consumed, the central universe could conceivably start a big collapse. Development of the mentioned mathematical model is left to Nancy's cousin Wm Clifton Bean, or some other mathematical genius. As speculated in a previous question/answer, if there are more than one universes, there could be an equivalent number of big squeezes. Like in fog, mist, or clouds, the universe could fragment and gravity could cause a multiple of squeezes, in differing directions.

Q *If the deflagration front runs out of fuel will the event be observable?*

A Possibly. But first we must develop telescopes and technology that will let us see the deflagration front. Since the Stage II reduction mechanisms transpire over a period of 2.5, and maybe as much as 5 billion years, the current nearest distance to the trailing edge of the front is at least 14+ billion years. It could have happened 16.5 billion years ago, but the light from the event hasn't reached us yet.!

Q *What initiated the deflagration front?*

A Statistical probability theories indicate that it more than likely was a rare spontaneous event. In other words, Professor Dr Alvin Pershing at Oklahoma State University broke his chalk one too many times. Or a minute electron volt electrical spark occurred from who knows where. In other words, the answer to this question is not known.

Q *Frequently asked is the above question, "What initiated this process?"*

A. In the late 1940's through the 1960's, Professor Dr. Alvin Pershing taught "Quantum Mechanics", "Thermodynamics", "Kinetic Theory of Gasses",

"Line Spectra" among other subjects. His classroom had blackboards on three walls and the other wall was all windows. Most of his courses were heavily involved with statistical analysis. Every day from the starting bell he would promptly start talking and writing, filling the blackboard with equations, just as rapidly as he progressed clockwise around the room. As soon as he came around, back to the starting place, he would pick up an erasure and erase with one hand and write with the other, all while giving his technically entrancing lecture. Occasionally he would get started on his third lap before the ending bell. Needless to say he wrote fast and used up large quantities of chalk. Periodically he injected a little humor; when the chalk would break he quickly snapped his head around and looked up to the ceiling. I was the sucker early in my first semester in his class and asked the question why? He said he had worked out statistically that once in every few trillion years the chalk would fall up rather than down. If it happened to him he did not want to miss the rare opportunity to witness the event !

Speculation;
..........The postulated composition and density of the proto-matter outside of the deflagration front is rarified and the statistical probability of even just one annihilation would surely be rare. It may have been caused by a stray photon, heavy neutrino, or a rogue gravity wave if there is such a thing. Maybe the same rogue gravity wave would someday cause Professor Alvin Pershing's chalk to fall upward, if he were still alive! But it only needed to happen once, and we are glad it did, at least the once of which we are aware.

Q   ***What if the Hubble numbers are, after due analyses,*** not inconsistent because of direction of measurement, what happens to the New Universe Theory center of the universe as theorized in this document.

A Then the center of the universe is elsewhere and additional analyses is appropriate. As stated in the text, the graphical analyses, as presented, is based on what we believe are competently measured H number data. If these data are proven to be invalid, then the center of the universe may be closer or farther than my estimated value of 3.5 BLY.

Q *There are several combinations of galaxies that* appear to be in the same neighborhood, yet they have significantly differing red shifts. These discordant red-shift observations appear contradictory to either the BB or the NUT. How do you answer this?

A The discordant red-shift galaxies could be in the same neighborhood because the two are members of different super-galaxy vortexing structures and are on opposite sides of their huge vortexes. Another scenario is that there may be as suggested in another question, more than one deflagration front and are simply traveling and decelerating in somewhat different directions. (Not likely or discordant would be more common)

Q The graph of Figure 4.3 shows the scatter band of H number vs year is narrowing with the passing of time. What gives?

A Hubble number calculations are deduced from red shift measurements compared to various distance candles. The convergence of the numbers indicates that astronomers are in recent years using the same objects for their reference and for their calculations. That is why we need a catalogue of H numbers and directions, as well as reference distance candles from the same vicinity where the red-shift data was obtained. I don't accept for a minute that the earlier astronomers were incompetent, but their instruments were not as precise as is currently available. Also, their assumption that Hubble numbers should be the same in all directions is erroneous.

Q   *You refer to the universe as spherical. Why does it* have to be spherical?

A   It doesn't. It could be any three dimensional shape, and there is a reason to suspect that it may be a twelve sided polyhedron, or a six sided hexagon. Similar to the geometry of the primordial crystalline lattice.

Q   *Since mass cannot travel as fast as the speed of light* as explained by Einstein's Laws of Relativity, how fast can it travel? The deflagration front must be traveling away from the center of the universe at the maximum. Can we determine what is the speed of this "potato light truck"?

A   Possibly. Since we know the annihilation front is radiating Gamma Rays (electromagnetic waves in the shorter than single digit Angstrom unit wave lengths), and since we know the spectral characteristics of positron annihilation radiation from the anti-matter above the milky way galaxy. All of the radiation energy from the annihilation front is surely not being consumed by transformation back into mass, and since some, if not most, of the COBE or WMAP measured background radiation is 'left over energy' from the deflagration front. The wave lengths would allow us to calculate the red-shift, from which we can calculate the difference in speed between us and the "front". (Since we do not yet totally understand the annihilation processes, future observations and analyses results may address or unveil currently unknown questions and processes).

Q   There has been a lot of talk about so many hydrogen atoms being the dominant nuclide observed in the universe (Some experts have stated 65+%), why weren't just as many or more free multiple unit neutron nuclides produced by the reduction mechanisms as there are multiple unit proton neutron combination isotopes?

Bobby McGehee

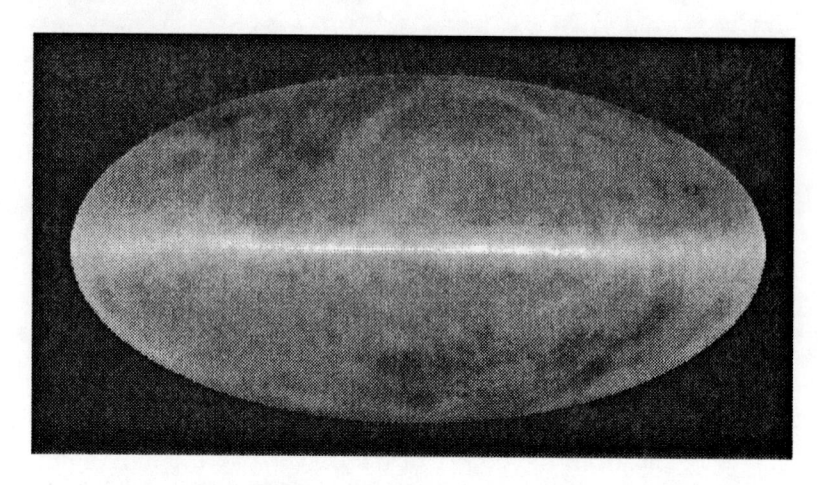

Figure 11.1 Sky Full Of Hydrogen Interstellar space is filled with extremely tenuous clouds of gas which are mostly Hydrogen. Radio Telescopes have mapped this image which represents an all-sky Hydrogen survey. The plane of out milky Way galaxy runs through the center. This picture with no visible stars, was downloaded from APOD December 18, 1996 and credit is to J Dickey (U Mn), F. Lockman (NRAO).

A　　They were. Evidence indicates there are many more (maybe in excess of 6 times more free multiple neutron mass than all other mass, see ref 42) multiple neutron nuclides than the total of all other mass particles. This indicates the possibility that much, if not all, of the unseen mass (gravitational source) that is known to exist because of the galaxy rotation speeds and stability, is not dark matter at all. It could simply be transparent matter (compound neutron nuclides). Single neutron nuclides are unstable and have a rest half life of only 10.25 minutes (see Figure 6.10). Multiple unit neutron nuclide isotopes are apparently extremely stable, as they or their decay processes have not yet been observed. Multiple neutron nuclides are weakly interacting mass particles, (WIMPS) and about the only way they can be detected is by observing their gravity effects. Guess what? That is exactly what we have been doing. Gravity influence has been telling us the answer, not just the question. Kepler's laws tell us that there is more mass (about 6 to 10 times more) than we have been able to otherwise detect. **Another case of "Open your eyes and look at me"!**

Q　　*If "free" neutrons are so short lived, why can't their be long lived pure multiple neutron Nuclide isotopes?*

A　　The half-life of individual neutron nuclides is 10.25 minutes according to the most recent (2002) Nuclide data published by the Knolls Atomic Power Laboratory. (Half-life means that after 10.25 minutes one half of any given quantity of the free neutrons will have disintegrated into a proton + a beta particle (electron) + a neutrino. However, since atomic mass neutron nuclides of 2 or more have never been detected, they may have the longest of half-lives and that may be why they have never been detected, except by gravitational interactions. They could be the major part of the mass of the universe! As a matter of fact, that is one of the reasons I included the lower segment of the "NUCLIDE" chart in Chapter 6. If you examine and scrutinize the zero proton row (bottom row) in the nuclide chart it raises the same question. Original nuclides were formed by coalescing protons and neutrons caused by extremely high velocity-enhanced gravity forces. It appears highly likely that there were as many, probably many more, multiple neutron-nucleus-isotopes than isotopes of all other combinations of neutrons and protons. Therefore, apparently there may be many times more free pure neutron isotopes existing in the universe than; Hydrogen, Helium, Lithium, Beryllium, Boron, and maybe even carbon isotopes. While we are speculating, the foregoing statement could indicate that much as 90% of the universe's mass is made up of free neutron isotopes. <u>Dark matter is simply Transparent Matter that does not interact except by gravity!.</u>

Q ***How does this New Universe Theory for generation*** of the universe differ from the "Continuous Creation Theory"?

A This new Universe Theory is solely invented/discovered by myself, and is not a rejuvenation of the Thomas Gold continuous creation (steady state) idea. It is far from that concept that was originated and envisioned by the Austrian-born astronomer Thomas Gold. His concept theorized continuous reformation of mass everywhere in the universe to offset the mass being consumed and dissipated as energy by the star 'processes'. Gold's concept assumed spontaneous and continuous 'creation' of mass particles uniformly throughout the universe. The most outspoken supporter was Fred Hoyle. However, defeat of that concept was conceded with the discovery of quasars. George Gamow conceded early on to the BB. Fred Hoyle stood his ground, respecting the Laws of Physics, and continued to do so to his death in 2001, as also did Grote Reber, who died early in 2002. They were not specifically in support of the Gold continuous creation theory, but Hoyle and Reber both refused to accept any concept that violates the Laws of Physics. The New Universe Theory is of continuous generation or transformation, not continuous creation. The Thomas Gold concept was only in a few ways, somewhat similar to this technically feasible New Universe Theory. Fred Hoyle, Grote Reber, visionaries among many others, maintained credibility and could not accept the idea of everything coming from nothing at a single point. Hoyle was often ridiculed for not accepting the BB, but this New Universe Theory now redeems him for his steadfast integrity. From what I have read about Fred Hoyle and Grote Reber, I believe if they were still with us, they would be gentlemen and would not "rub it in" to past dissenters, even though some were insulting to them. I will do the same for the University of Wichita vector analysis math instructor, by not mentioning his name.

Q **Doesn't this New Universe Theory eliminate the** purpose or need for a God? (A question by a young bewildered student).

A To answer that far reaching question, with your indulgence, I wish to repeat a parable I once heard: *One day, there were two brilliant well educated physicians talking to each other and one said to the other, 'You know, we now have unraveled genetics* *and all of the secrets of life, so now we can create a man or any other kind of a living being from a shovel full of dirt.' The other said 'That is right, so we don't need God anymore.' Immediately a booming voice came from the sky and the voice said; 'If you really can create anything at all, please show me.' The first physician said; 'O.K., just watch this'. As he scooped up a shovel full of dirt, the booming voice again rang out; 'hey there, get your own dirt'!*

If the physicians had been a scientific thinkers, they would know the first law of physics always applies; "Matter can neither created nor destroyed."

Q *What do you think was/is the origin of all the matter in the universe?*

A This is a question for further thought. Maybe someone in the future can answer this one.

Q *Many questions other than the ones suggested will* come to mind. Where can other inquires be presented for comment?

A Questions and suggested answers will be accepted on the web site WWW/NewUniverseTheoryQuestions.com.

Q *If all the so called "missing mass" is simply the undetectable (except for gravity) elusive neutron nuclides, doesn't that imply that the primordial matter density must also have more density?*

A Precisely. The continuity laws demand it. The spatial separation of positroniums may be much less than speculated in the Stage I primordial matter descriptions. But that does not change the "Reduction Mechanism" processes.

Q. *Can we directly monitor, over a period of time, the red-shift of single, specific distant galaxies or quasars and thereby directly calibrate the deceleration rate of objects during the early development of our Stage III universe?*

A. Deceleration does not occur on individual objects. The reduction in speed of groups of objects is a result of interactions and transfer of the linear velocity to revolving about another object, so the individual object does not slow, only changes direction and does so, on an average, over the period of revolutions about other object(s). Deceleration in our region of the universe comes from interaction of large assemblies, such as super-galaxies. The linear momentum transfer occurs from the tangent speed at the periphery of the super-galaxy. But the individual objects speeds are not changed in magnitude, only from linear motion to rotational. The change in transition speeds occur at the larger scale of the super-galaxy. Back to the question; The rotational speed is high, but the distance is so great, measurement of change would require observation over a period of several hundred million years to measure individual galaxy transitions from linear to orbiting (over a several degree segment of their newly acquired orbit). And at the distance to the beginning edge of Stage III where only individual galaxy interactions are occurring, the distance is excessive. We have not yet observed the motion of anything at that distance other than incremental velocity of quasars by red shift. The answer is...The universe's deceleration cannot be directly measured. Astronomers are known for their perseverance, but the answer to the question is, "No, I don't think so".

Q *What would Sir Fred Hoyle, Grote Reber, Thomas Gold, and the thousands of deceased others that never bought into the BB say, if they were here today?.*

A **Sir Fred Hoyle** and **Grote Reber** deserve this revelation, they had the credibility I wish for all scientists, present and future. (See Tributes ...Appendix 8.0) True scientists are forever students and do not limit their learning by gloating about previously achieved credentials. All new knowledge, should stimulate more questions than were answered by the newly acquired information. There are many questions for the competent and capable professionals, some obvious, some subtle, mostly yet unknown. I expect professional and amateur cosmologists and astronomers as well as other scientists will respond positively, to this document, and pursue answers and further questions relative to the new Universe Theory. From my brief college teaching experience I found the most profound (penetrating deeply into the subject of thought) and most thought provoking and difficult to answer questions came from bright eyed inquiring minded students. If you have a legitimate question, please let it be known.

Q *Why was this New Universe Theory printed using a "self publishing" company rather than through a scholarly (University) publisher, or through a recognized scientific technical society?*

A Although their required peer reviews are highly desired, I felt the need to protect the confidentiality of my theory until it was actually in publication and available to all interested persons. I feared that reviewers would not be able to resist 'breaking the news' before the NUT was published.

Chapter 12
Conclusions

New knowledge and Epilog:

1. The **New Universe Theory (NUT)**; of all 'origin of the universe' concepts, This is the only known scientifically logical and **Laws of Physics** compliant answer to the question; "What is the source of matter and what physical processes build the Universe?".

2. The **"Open Your Eyes"** new interpretation of increasing red-shift with distance for astronomical observations negates the Big Bang hypothesis. The fact that red-shift increases with distance was erroneously interpreted as acceleration, and that has misled astronomers for more than 75 years. (Until now).

3. The Big Bang (BB) has been shown to be an invalid hypothesis. It never qualified as a scientific theory. Many hypothetical crutch/concepts have been invented over the past several decades in an attempt to explain away Laws of Physics violations by the hypothetical BB, and to otherwise support the BB hypothesis. These crutches are negated by the NUT. Specific examples:

a. Inflation.

   b. Theoretical past existence of Population III stars.

   c. Eras of variable inflation.

   d. Hypothesis about variations in the speed of light in the early universe.

   e. The unacceptable idea that the Laws of Physics may have been different in the early universe.

   f. Ideas of multiple dimensions for the source of the universe.

   g. Dark energy hypothesis for explaining the illusion of a an exploding and accelerating universe.

   h. Strings and rubber band concepts.

4. The **New Universe Theory** describes development of the universe through three stages: **Stage I** is the primordial matter. **Stage II** is the transformation region which is described as a deflagration front (DF), wherein the transition of matter occurs in three overlapping sub-Stages. Graphical analyses reveal the deflagration processes are parts of the NUT shell that is approximately 2.5 BLY thick. It is traversing through space at the speed of light, continuously increasing the size of the universe in all directions. **Stage III** is the contained volume of matter. It includes all potentially observable (directly and indirectly) objects and energy.

5. The **NUT** presents a plausible concept for the hidden mass that provides the balance of gravity and centrifugal forces in the vicinity of galaxies and galactic clusters. Astronomer's pre-supposed idea of the existence of WIMPS (Weakly Interacting Massive Particles) is supported by the NUT. The

NUT provides logical basis for belief of the existence of neutral multiple neutron nuclides. These nuclide particles are theorized to range from two to more than two neutron masses, and these particles are neutral, stable, non-interacting, and undetectable except by indirect observation of gravitational effects on other matter. Hypothetically, multiple neutron unit nuclide particles make up large invisible clouds which are held together by gravity and are integrated with, as well as forming halos around galaxies and globular clusters. These theorized particles do not chemically bond or accrete as they are electrostatically neutral. These neutral particles, individually and in clouds, contain considerably more mass than all other mass observed throughout the universe. Dark matter in galaxies is about 6 times all other mass, according to the most recent analyses (reported in "Science", 12 December 2003, Vol. 302 page 1902, by Ken C Freeman. His title of this scientific paper is; "The Hunt for Dark Matter in Galaxies". Ken Freeman is at Mount Stromolo Observatory, Australia National University, Weston Creek, ACT 2611, Australia).

6. Many physicists and cosmologists have the pre-supposed idea that electrical charges of electrons and positrons (matter and antimatter) should be equal within the aggregate universe. The NUT indicates that their idea is probably valid, as the quantities of charges are balanced in primordial matter. The Stage II process and the continuity laws preserve the balance. Also, decaying neutrons produce equal quantities of + and - particles.

7. Gravitational lenses as currently theorized may not be directly due to gravity. The observed refraction of light from distant objects as their light passes near galaxies and clusters of galaxies might be mostly caused by transparent lenses made of huge clouds of multiple neutron nuclides. The cloud density gradients and shape, and locations of these clouds are as a result of gravity from galaxies and super galaxies.

Only a portion of "gravitational lenses" are caused directly by gravity.

8. The conclusion and correct assessment of the Big Bang is; The BB was simply a presumptive conclusion from reverse extrapolation of mis-interpreted red-shift data. As pioneer radio astronomer Grote Reber often stated, the Big Bang is Bunk.

However, we must recognize astronomers and physicists that accepted the BB are limited by then available red-shift interpretations. I admire all individuals with the curiosity to search for validity in existing and new theories.

9. It is speculated from logic and proven physical fluid flow phenomena, there may be a region of void at the site of the DF initiation that is producing the universe. It is theorized to be centered at only 3.8 BLY away, and in the opposite direction from the Abell galactic clusters.

10. The universe is theorized as continuing to grow, at the speed of light, beyond our limited range of view. It is estimated to <u>currently</u> be a huge sphere with twice the radius of the most distant observable object, plus the distance back to the origin of the DF from the MWG. The universe's size is increasing while the distant object's light is progressing back to us, and at the same rate.

11. The universe is more than 78.8 BLY in diameter. From our Milky Way Galaxy (MWG) perspective the maximum apparent observable distances are between 13.4 and 21.0 BLY, dependent upon direction of observation. This coincides with the present theorized maximum observable red-shift, or 'red limit'. Among other amazing conclusions, this currently <u>observable</u> universe is 34.4 Billion Light Years

(BLY) in <u>diameter</u>, and we are ~3.8 BLY from the center of the universe.

12. Hubble lines were reconciled (Figure 10.1) from initial tentative straight line graphing (Figure 7.6) to match Richer's proof of the minimum age of our region of the universe. With the most distant observed objects, the Hubble lines now need to be further reconciled to the potentially observable distance, at the boundary between the Stage III and Stage II in the DF. The curvature of the Hubble lines have been only estimated to date. At velocities below 150,000 kilometers per second the Hubble lines are straight. These have been calibrated by correlation of red-shifts and distances confirmed by comparison of brightness and the apparent brightness of Type la supernovae. Yet to be done math modeling of the higher velocity region should allow better graphing for Hubble line reconciliation. The, yet to be refined, new Hubble line extensions are expected to reveal that the universe diameter to be two, possibly three times the initial (based on Richer data) reconciled estimates. With new reconciled lines the observable universe diameter could prove to be over 100 BLY, and the diameter of the physical universe could be 200 BLY or more.

Bobby McGehee

Figure 12.1. Far-Young Objects. This the photograph displays the results of an extremely long exposure time (during a period of 3 months staring at the same spot) towards a small segment of a the shell of our universe. This was accomplished using the Hubble telescope, continuously aiming the Hubble at that small spot while the hubble revolved around earth and while earth revolved around sol. NASA's release of this photograph in March 2004 is very timely, to coincide with the finalizing of this manuscript for submission to the publisher. It is recommended that the reader view a full color image of the above figure by going on the internet at: http:antwrp.gsfc.nasa.gov/apod/apo40309. html . This web page is also found by searching for APOD (Astronomy Picture Of the Day) and from the archive locate 2004 March 9. Click on the photo for the highest resolution available. (Very impressive, Thanks NASA.)

13. A deep sky photo taken using the Hubble Telescope reveals "Uncharted Depths" in the words as used in the 12 March 2004 publication of the AAAS, page 1596, Volume 303. Our human eyes are sensitive to light in the wave length range from ultraviolet to infrared, (which corresponds to light wave lengths from 3200 $\underline{A}$ to 7700 $\underline{A}$). The youngest and more distant stars are the hottest burning, emitting light at the blue end of the spectrum. If we assume the stars in this photo emitted light at the blue end of the visible spectrum and the light from the farthest and fastest objects is velocity shifted to the red end of the human observable light. The shift (can be calculated; shift from 3200 $\underline{A}$ *(Angstrom units = 10^{-8} centimeters)* to 7700 $\underline{A}$ divided by 3200, minus 1.0 results in red-shift, z value of 1.406. (Equations are described in Appendix 3.0) With the NUT developed graph (Figure 10.1) we can deduct the velocity of the reddest observed object as about 6 tenths the speed of light.

If the picture was taken in the same direction as the Abell clusters, it would be in the direction of the closest proximity of the deflagration front, and the steepest Hubble curve of 117 could be used to graphically determine the object's distance to be 6 BLY away from us at the time it emitted the light we are observing here and now as red. However, the picture was taken in the direction of constellation Fornax, which is almost opposite the Abell clusters. One significance of this is the photo of *Figure 12.1, Older and Younger Objects, was taken in the direction back towards the origin of the deflagration front (and the starting place for the universe's growth). Some of the objects in this view are before our region existed and, at greater distances beyond about 6 or 7 billion light years, objects are younger, as in the opposite direction..*

With this photographic data allowing such extensions of our comprehension of our universe, just think what might be revealed with similar exposures of distant objects if the exposure times were doubled or quadrupled. And then do it again in other directions.

Other, newer technology space telescopes could do this job better than the Hubble Space Telescope (HST), but their time is too valuable to commit them to that much time to this project. The HST could be committed to such along time consuming projects, instead of committing it to extinction. A project worth considering.

Special Recommendation:

In the March 12, 2004 issue of the AAAS Journal of "Science", page 1596, Robert Irion reports about the Hubble Space Telescope image released March 9, 2004. He quotes Steven Beckwith, director of Space Telescope Science Institute (STScI), at Baltimore Maryland; "...... the quality of data ..... is better than anything ever done with the space telescope." Backwith conceived the "Ultra Deep Field" (UDF) project which included the 100 day long time-exposure of remote galaxies. They produced a detailed, sharp, image of about 10,000 very far away galaxies while viewing a tiny patch of sky in the Fornax constellation.

This and many other projects are proving to be very productive and useful. Some of the projects may not be of great scientific value today, but if they provide comprehension and enjoyment for we taxpayers, I say lets do them. It is recommended that NASA management re-structure their budgets to allow the HST to continue operations until a significant length of time after the newer replacement space telescopes are actually in operation and are proven successful. In due time the scheduled demise of the HST will be proper, but I believe it is premature at this time. Scheduling of research projects that use HST should be allowed to continue.

NEW UNIVERSE THEORY WITH THE LAWS OF PHYSICS

Both the March 12, 2004 "Science" issue reporter and the UDF project manager refer to the BB in their comments, however now, in the light of the New Universe Theory (NUT), they can re-interpret their observations. The acquired knowledge with use of the HST is even more valuable and revealing than previously thought. Let's keep the HST operating for at least another decade (to 2020 or 2030). In due time, it can then be allowed to die with the full dignity and honor it deserves, but premature irreversible action, or lack of action, could prove to be a dire mistake. Present schedulers will be thanked for generations to come.

More Far-Out Thoughts:

It is comforting and refreshing to know that we the occupants of the universe, including the many other probable (some hopefully even more intelligent) biological communities, will have a home and will continue to exist for a long, long time, by anyone's measure. All star systems and each biological community, as proven on earth by observation of all species, have limited time spans for their existence. However, it is probable that other biological intelligent species have, and more will emerge. As stated in the synopsis and purpose of this NUT book, the universe is not expanding to its death, it is growing with vim and vigor!

We are providing at least one conscious intelligence for the universe. We may never know if we are alone. Therefore, long before our solar system inflates into a planetary nebula, we hopefully will have developed super intelligent robot species that can improve and build more of their kind, that aren't as fragile as we biological beings. Such technological potential will be feasible within a few decades. Our intelligent robotic self propagating 'descendants' can be dispersed to other solar systems, maybe even beyond the MWG, thereby preserving and perpetuating the intelligent consciousness that was biologically acquired here on earth.

<u>Epilog:</u>

Hopefully, the New Universe Theory will put the Astronomy and Physics community back on track. May thinking civilization never again deviate so far from the Laws of Physics. All of mankind's knowledge is built by adding new found facts onto proven and physically possible phenomena. Old comfortable ideas and myths from our beloved ancestors need to be frequently reexamined, and some concepts need to be released and cast aside when they are superceded by fact based and scientifically feasible answers.

Every time the statement is made or implied that we now know all the answers except for details, a lot of embarrassment is soon to follow. A recent publication (November 2004 "Physics Today") suggested that everything is now understood! Forward thinking mentors talk about what they don't know, not about how much they do know. Paraphrasing Stephen Hawking; For anyone to imply that with adequate time, they could explain the universe, is absurd! About a century ago one of the then US presidents suggested that the Patent office could soon be closed since everything had already been invented!

The New Universe Theory gives us a new perspective, answers many questions, and solves many mysteries. But, as stated in Chapter 4.0, it opens the door to an endless list of more exciting phenomena to research. There are so many things that we don't yet know that we don't know.

At this time, it is most urgent that the New Universe Theory be revealed so it can be kibitzed and used. Refinement will be forthcoming from highly qualified Cosmologists, Physicists and Astronomers. The solution of many (uncountable) riddles and puzzles remain for the students, amateurs, mentors, experts, and also for text book authors.

Glossary

Terms and Definitions

Semantic/s......relating to meanings of words; /the study of meanings in language.

Conveying information for understanding requires proper use and interpretation of terms. Misuse of terms indicates that in-depth understanding of the subject may be incomplete or faulty. Many terms have multiple definitions, so the meanings as intended in this book are provided with terse definitions to assure the intended meaning is understood. These terms and others should be occasionally perused and reviewed when considering this New Universe Theory, or reviewing any previously 'accepted' theory. Scientific thinking also requires assurance that all facts for logical deductions are valid and not someone's assumption or hypothetical conjecture. Scrutinizing with scepticism is one mark of a true scientist.

(Definition excerpts are composed from references such as: "Webster's 1996 Encyclopedic Dictionary, Random House Publishing"; and "Exploring the Dynamic Universe, 1995, West Publishing", and others, listed in references.)

Abscissa ...   Coordinate of a point from the y-axis measured along a line parallel to the x-axis.

Accretion ...   Growth,....Increase by external addition, ...A growing together of parts that are naturally separate; ..... The result of such growth or accumulation.

Adhere ...   To be consistent with, ... to agree,... to be attached or to be devoted...

Adventure ...   Bold undertaking in which unforseen experiences are to be met.

Amateur ...   One who engages in art, science, sport, or other activity for the enjoyment rather than for the money....

Analysis ...   The separating of any material or abstract entity into its constituents for study of those components.....

Anisotropy ...   Unequal physical properties along different axes.

Angstrom ...   A minute unit of length equal to one hundred millionth of a centimeter...(named for A J Angstrom, Sw. physicist)

Annihilation ...   Process in which a particle and an antiparticle unite, obliterate each other, and produce photons. The conversion of rest mass into energy in the form of electromagnetic radiation.

Ardor ... Great enthusiasm, ... eagerness, ...

Ascension ... See *Right Ascension.*

Astound ... To stun with bewildering wonder.

Atom ... The smallest component of an element consisting of a nucleus containing combinations of neutrons and protons and one or more electrons bound to the nucleus by electrical attraction. (The number of protons determines the identity of the element)

Atomic Number ... The number of positive charges or protons in the nucleus of an atom of any specific element.

Audit ... A scrupulous examination. An accounting as adjusted by auditors (those qualified to scrutinize).

Autumnal Equinox . The occasion when the sun crosses the celestial equator from north to south.

Axis ... A straight line, real or imaginary, passing through a body that actually or supposedly revolves upon it... A line passing through a body or system around which the parts are symmetrically arranged .

Baryons ... A class of elementary particles including protons, neutrons, and some other heavier particles.

Behemoth ...　　Something of monstrous size and power.

Belief ...　　Acceptance of something as fact even \if it is without substantiation and valid proof (see faith)

Bias ...　　An inclination that prevents unprejudiced consideration of a question...

Bibliography ...　　A list of works consulted by an author in the preparation of a book.

Big-Bang ...　　"A term referring to any theory (hypothesis) of cosmology in which the universe began at a single point, was very hot initially, and has expanded from that state since."

Big-Bang Hypothesis　　An idea that "caught on" because of the lack of a scientific explanation of what now is shown was an incorrect interpretation of red-shift of distant observed galaxies. It was mis-named a theory.

Big-Bang Theory...　　"A theory that deduces a cataclysmic birth of the universe from the (apparently) 'observed' expansion of the universe, cosmic background radiation, abundance of elements, and the laws of physics." (This definition is quoted from Reference 1.) (The writers and editors of that encyclopedic dictionary have mistakenly used the phrase 'and agrees' with the laws of physics). (Also it was a hypothesized concept, not a theory)

Bigotry...   Stubborn, complete intolerance of any belief, or opinion that differs from one's own.

Blasphemer ...   One who speaks irreverently orimpiously of sacred things ... or God.

Blather ...   Babble or foolish talk

Bold ...   Daring,..Presumptuous,..Audacious,.. Spirited, ...Adventurous

Calculate ...   To reckon by exercise of practical judgement. Estimating the probability of success. Also to determine by mathematical processes, including graphical.

Calibrate ...   To adjust and correct the divisions or graduations of a measuring device or system.

Causality....   Relation of cause and effect.

Celerity ...   Swiftness, Speed.

Celestial Equator ...   An imaginary circle formed by the intersection of the earth's equatorial plane and the celestial sphere. This is the reference from which an angle of declination is measured.

Celestial Pole ...   The line passes perpendicularly through the earth's equator and the celestial equator. The celestial line which corresponds to the earth's north-south pole line.

Celestial Sphere ... Pertaining to the sky or visible heavens, ... a sphere of indefinitely great radius.

Cluster ... A number of similar things growing or collecting together..

Constellation ... A prominent pattern of bright stars, historically associated with mythical figures. In modern usage each constellation incorporates a precisely region of the sky.

Concept ... An idea of something formed bycombining all its characteristics or particulars....

Conflagration ... A destructive fire, usually an extensive one (from outside).

Conjecture ... Inference from presumptive evidence... a guess or theory based on such evidence..sometimes from defective evidence...

Coordinate ... Any of a set of numbers that gives the location of a point.

Corpuscle ... A minute elementary particle of matter, usually thought of as a minute elementary particle of mass.

Cosmology ... A term that is technically, the study of the universe as it now appears is cosmology. The study of its origins is cosmogony, and this term applies to the big-bang as well as the earlier theories on the origins. Cosmology is generally used for both.

Couple...	A pair of equal parallel forces acting in opposite directions tending to produce rotation. Rotation of a couple can only be stopped by another couple.
Credible...	Worthy of belief or confidence; trustworthy
Creation ...	The act of causing to exist. The act of constituting.
Crystal Lattice ...	The repeating symmetrical pattern... that occurs in a crystal
Crystalline Solid ...	A solid in which the particles are arranged in an orderly repeating pattern.
Crystal ...	A solid body enclosed by symmetrically arranged plane surfaces, intersecting at definite characteristic angles, and having the same characteristic structure throughout.
Crystals, covalent ...	Crystal with a network of bonds that extend throughout the entire solid.
Declination ...	The angular distance, in degrees, north or south of the celestial equator, which is the projection of the earth's equator into the sky.
Deduction ...	A process of reasoning in which a conclusion follows necessarily......
Deflagration ...	To burn, especially suddenly and violently (from inside out)

Density ...	The quantity of any substance per unit volume
Didactic ...	Intended for instruction.....intending to teach a moral lesson.
Dilatation ...	Spread out, broaden, or expand. For time it is an increase in time per unit of time.
Dilemma ...	A situation requiring a selection of preference between unpleasant alternatives.
Dogmatism ...	Arrogant assertion of opinions as truths, ...unfounded positiveness...
Element ...	A substance that cannot be separated into simpler components by chemical processes.
Elemental ...	Pertaining to a simple, basic, and ncompounded constituent
Emit ...	To send out, ...to send forth, ...to throw off....
Energy ...	..(physics) The capacity to do work. A constituent of matter.
Enlighten ...	To shed the light of truth and knowledge upon...To free from ignorance, error, ..To inform...
Enthalpy...	The internal energy of a system. ...also called heat content or total heat..

Entire ...	Complete, The whole thing, Having no part left out.
Entropy ...	(thermodynamics) A measure of the amount of energy that is not available for work. .... (statistical mechanics)...A measure of randomness of .... constituents. A closed system evolves toward a state of maximum entropy......(cosmology) ...a hypothetical tendency for the universe to attain a state of maximum homogeneity....
Erudite ...	Widely knowledgeable mostly from book study.
Erudition ...	Knowledge acquired by study, research,.....an instruction...
Evidence ...	That which tends to prove or disprove something...ground for belief..
Fact ...	A truth known by actual experience or observation. Something known to be true.
Faith ...	Belief that is not based on proof........
Fictitious ...	Assumed for the sake of concealment, not genuine...
Fantasized ...	Dreamt of or hoped for; longingly imagined,; fancied......
Fizzle ...	To fail after a good start.

Galactic Cluster ...   A loose cluster of stars located in thedisk or spiral arms of the galaxy. Also referred to as **Open Cluster.**

Geometry ...   That branch of mathematics which investigates the relations, properties, and measurements of solids, surface, lines and angles. The study of space and of figures in space.

Globular ...   Globe shaped.

Globular Cluster ...   A large spherical cluster of stars located in the halo of the galaxy. These clusters, containing up to several hundred thousand members are thought to be the oldest objects in the galaxy.

Gravity ...   The natural force of attraction between any two massive bodies.

Hadrons ...   Any of several elementary particles; protons, neutrons, and mesons.....

Hallucination ...   Something that is imagined and exists only in the mind

Hexahedron ...   A six sided solid figure. Usually opposite sides are parallel planes.

Homogeneous ...   Of uniform make-up or structure.

Hypotheses ...   A proposition assumed and set forth as an explanation for some specified phenomena,..asserted as a provisional conjecture,...

Hysteria ...	An uncontrollable outburst of emotion ,...characterized by irrationality
Idea ...	Any conception existing in the mind as a result of mental understanding, or awareness...
Incipient ...	On the verge or edge of existence...
Infamy ...	Very bad reputation,.. Public disgrace,... Shameful...
Infinity ...	Unlimited quantity of anything; space, time, mass, energy, matter; boundless; an indefinite great number or amount.
Intrigue ...	To arouse the curiosity or interest
Intuition ...	Apprehension, ... cognition,... ..the power of knowing without recourse to inference,..... insight.
Invention ...	Act of finding out; discovery. The power of conceiving, devising, originating out of ingenuity.
Inviolable ...	Secure from infringement, corruption, or destruction.
Isothermal ...	Indicating changes in pressure and volume at constant temperature.
Isotope ...	One or more atoms that occupy the same position in the periodic table, identical in chemical behavior, and distinguishable only in radioactive

	transformations, and with neutron mass incremental differences in atomic weight.
Isotropic ...	Having the same properties in all directions.
Kinetic ...	Of or having to do with motion,... Caused by or resulting from motion.
Know ...	To apprehend clearly and with certainty..to perceive or understand as fact or truth....
Knowledge ...	Acquaintance with facts, truths, or principals, as from study or investigation...
Lagrange ...	Orbit locations (first discovered by J L Lagrange) around a celestial body that are stable. These are 60 degrees leading or 60 degrees following another orbiting body. For example, the moon.
Lattice ...	A three dimensional array or a repetitive pattern that describes the long range order and arrangement of particles in a crystalline solid...
Law ...	A statement of a relation or sequence of physical phenomena, repeatable under the same conditions.....
Leptons ...	A class of fundamental, elementary particles that are believed to be truly elementary and not made of other

subunits. Low mass particles and their anti-particles.

Locus ...   **All** points on a curve that describe a mathematical equation.

Logic ...   The science that investigates the principals governing correct or reliable inference,......the system or principals of reasoning applicable to any branch of knowledge or study

Mass ...   A body of coherent matter......a quantity of matter as determined from its weight.....a constituent of matter.

Matter ...   The substance or substances of which any physical object consists or is composed..., **Matter consists of both mass and energy...**

Mechanisms ...   The doctrine that natural processes are mechanically determined and capable of explanation by the laws of physics and chemistry.

MegaParsec ...   **One** million parsecs. (See parsec)

Miracle ...   An extraordinary event that surpasses all known human or natural powers ....(usually an unrepeatable and therefore unprovable, or unexplainable event)

Monolithic ...   Made of only one piece, ...Unbroken, ...Total uniformity and rigidity.

Myth ...	Any invented story, idea, or concept,.... an imaginary or fictitious thing or person...
Negate ...	To prove as nonexistent and, or to nullify.
Neutron ...	An elementary particle having no charge, with a mass slightly greater than a proton. (It decays into a protron, an electron, and a neutrino)
Neutron Star ...	A celestial body of super-dense remains of a star that has collapsed with sufficient force to pull all of its electrons into the nuclei that they orbit, thus leaving only neutrons.
Novice ...	A beginner. A person who is new to the circumstances in which he or she is placed.
Nomograph ...	Graph from which numerical values or related variables can be deduced.
Nuclide ...	An atomic nucleus characterized by its atomic mass and atomic number. Any isotope without its orbiting electrons.
Objective ...	Not influenced by personal feelings, interpretations, or prejudice; based on facts.
Observation ...	An act or instance of viewing a fact or occurrence for some scientific purpose...

Ordinate ...   The coordinate representing the distance from a specific point to the x-axis, measured parallel to the y-axis.

Paradox ...   A subject that seems self contradictory or absurd, but upon close examination reveals in reality a possible truth.

Parsec ...   Unit of measure for interstellar space... equal to a distance having a heliocentric parallax of one second.... equal to 3.26 light years, ..or 19.2 trillion miles.

Peer ....   Another person of equal status, scientific or technical qualifications.

Periphery ...   The surface or outer edge of any body.

Phenomena ...   An occurrence that can be observed or perceived.

Photon ...   A quantity of electromagnetic radiation...usually considered an elementary particle that is its own antiparticle and has zero rest mass, and no charge. Also called a light quantum. An energy quantum that is a constituent of matter.

Physics ...   The science of mass and energy (Matter) and the interactions between them.

Planetesimal ...	A small (diameter up to several hundred kilometers) body of the type that first condensed from nebula.
Plausible ...	Believable, appearing true, reasonable.
Positron ...	An elementary particle having the same mass as an electron, but having a positive charge equal in magnitude as that in an electron. The antiparticle of the electron.
Positronium ...	A short lived subatomic system composed of an electron and a positron, held together by electrostatic charge, but separated by their mutual orbiting centrifugal forces.
Postulate ...	To assume without proof.
Precursory ...	Preliminary.
Prejudice ...	An opinion formed beforehand, without thought, knowledge, or reason.....any preconceived opinion or feeling, either favorable or unfavorable....
Premise ...	To set forth beforehand....as a basis for a conclusion...
Preposterous ...	Contrary to nature, reason, or common sense; absurd; senseless..

Primordial ...	First existing,... primordial matter..
Principle ...	A fundamental truth....a primary or basic law, doctrine,...
Proof ...	Evidence sufficient to establish a thing as true....the establishment of the truth....
Propagate ...	To cause an effect at a distance, as by electromagnetic waves, etc., traveling through space or a physical medium.
Proto- ...	A combining form meaning first... A substance which is held to be parent of the substance to which it is prefixed.
Pseudo ...	A combining form meaning fake, falsely, ... deceptively resemblance to a thing, ...false,... to deceive,....
Puzzle ...	To exercise one's mind to think through a problem, ... To bewilder or perplex.
Quark ...	A fundamental (theoretical) particle.. It is less than one third the size of a neutron or proton...(never, yet, observed in a laboratory)..
Quark Star ...	A theoretical celestial object that has collapsed with force sufficient to reduce all particles to strange quarks. Two stars have been found, in 2002 and 2003, that have the mass, size, and brightness to be quark star candidates.

Radiate ...	To emit rays, ... to proceed in a direct line or lines from...
Reconcile ...	To make agree, .. to bring into harmony,...
Redeem ...	To rescue or deliver, as from bondage,... to make amends...
Red Giant ...	A star of great size that has relatively low surface temperature.
Referee ...	An authority who evaluates scientific, technical, or scholarly papers, proposals, or the like for publication.
Resolve ...	To answer and explain, ..
Revelation ...	The disclosing to others what was before unknown to them... Often, a striking disclosure..
Riddle ...	A question that is puzzling and requires some ngenuity to arrive at an answer.
Right ascension ...	The directional position of an object in the sky as measured in hours, minutes, and seconds to the east from the vernal equinox, a fixed direction in space..
Sanctum ...	A sacred and inviolable place or retreat.

Science ...   A branch of knowledge or study of facts... systematic knowledge of the physical or material world gained through observation or experimentation

Scrupulous ...   Having and showing strict regard for what one considers principled and right.

Sidereal ...   As determined in reference to the stars, The primary geometric reference for the universe. Used for time and physical position reference. (A sidereal day is about 4 minutes shorter than a solar day).

Singularity ...   The mathematical representation of a black hole

Skeptic ...   A person who questions the validity or authenticity of something generally considered to be factual.

Speculation ...   The pondering of a subject in its different aspects and relations...to theorize from conjectures...

Startling ...   Causing sudden anxiety, surprise.

Sublimation ...   Transfer of a substance from the solid phase directly into the gas phase without passing through the liquid phase.

Subsidiary ...   Furnishing aid..aiding in an inferior status.

Substantiate ...   To establish by proof or competent evidence.

Superficial ....   Concerned only with the apparent or obvious. Not profound. Not genuine. Lacking substance.

Synopsis ...   A general view as of a treatise; .. condensed statement.

Technical ...   Having to do with an art, science, discipline, or profession.

Temperature ...   A measure of the warmness or coldness of an object with reference to some standard value. ....a measure of the mean value of the momentums of the particles making up the body of matter.

Tenuous ...   Not dense, .. rarefied,... not substantial..

Theory ...   A proposed explanation whose status is still conjectural....

Thought ...   A single act or product of thinking.... a judgement, opinion, or belief....

Transmuting ...   The changing from one nature, form, substance, or species into another.

Trapezium ...   A rectilinear (4 sided) quadrilateral plane projection of a figure with no parallel sides.

Truth ...   The actual state of a situation, .. conformity with fact or reality, ... a verified or indisputable fact, proposition, principal,.....

Universal ...   Applicable to all purposes, conditions, things, times, places, and situations.

Vernal Equinox ...   The occasion when the sun crosses the celestial equator from south to north.

Viable ...   Having a workable explanation and solution. Capable of growing and developing.

Vortex ...   A mass of ...spiraling inward or outward, having a whirling or circular motion...

White Dwaft ...   The remnant of a star that has collapsed, into an extremely dense state with no empty space between its atoms, but not reaching the extremely denser state of a neutron star or black hole.

Zealous ...   Passionate, ardor, ..intensely earnest, actively enthusiastic.

Appendix 1.0
"Laws of Physics"

The Laws of Physics, also, often referred to as The Laws of the Universe, are stated in the various scientific disciplines using slightly different phrases. These phenomena and principles have been tested, re-tested, and re-proven many times over, and therefore are referred to as Laws by the scientific community. The laws of physics are absolute and are not violable. They are not man made, they are only discovered by man. All are applicable to cosmology. Some are listed here, and have been abbreviated as they apply to the material of this document. Laws by definition, identify processes that are repeatable and verifiable by observation and experiment. Laws of Physics are universal.

Newton's First Law of Motion. (Law of classical mechanics)

>A body at rest or in a state of uniform motion, tends to stay at rest or in uniform motion unless an outside force acts upon it. (First observed by Galileo but Newton added the important notion of mass).

Newton's Second Law of motion.

The sum of the forces acting on a body is equal to the product of the mass of the body and the acceleration produced by the forces. (F = M A)

Third law of motion. (Also called Newton's law of motion).

For every force acting on a body, the body exerts a force of equal value and of opposite direction along the same line of action as the original force.

First Law of Thermodynamics. (Law of Conservation of Matter).

The principal is that; Matter can neither be created nor destroyed. The matter in any closed system is constant irrespective of its form.

Second Law of Thermodynamics.

In any process there is always an increase in Entropy.

Third Law of Thermodynamics.

Entropy is non-existent in a crystal at absolute zero temperature.

Newton's Law of Universal Gravitation.

Any two bodies in the universe are attracted to each other with a force that is proportional to the product of the masses of the two bodies and inversely proportional to the square of the distance between them.
$$F = k \times (M_1 \times M_2)/d^2$$

Kepler's First Law.

Each planet has an elliptical orbit with the center of mass between them and the sun at one focus. (Has since been extrapolated to apply to any two or more mutually orbiting objects, in or outside the solar system)

Kepler's Second Law.

Any object that rotates or moves around some center has angular momentum, which depends on its speed, its mass, and its distance from the center of motion. The angular momentum of an object is the product of its mass, its speed, and its distance from the center of mass.
(It is equal to the product of m X v X r)

Kepler's Third Law

The relationship between the period and semimajor axis of an orbit depends on the sum of the masses of the two objects.

Coulomb's Law.

The electrostatic attraction between two objects is proportional to the product of their electrostatic charges divided by the square of the distances between them.
$$F_q = (k \times (q_1 \times q_2)/d^2)$$

Einstein's Law for Transformation and the Conservation of Matter. (This is probably the most famous equation in history)

Energy is equal to the object's rest mass times the square
of the speed of light. **Constituents of Matter are mass and energy.**
$$(E = M_0 c^2)$$

Einstein's Law of the maximum speed of matter.

No object with mass can achieve the speed of light; An object's mass approaches infinity (not possible to achieve) as its speed approaches the speed of light.
$$M = M_0 \times (1 - (v^2/c^2))^{1/2}$$

Schwarzschild's Radius. (the radius from the center of a black hole to its event horizon)

The strength of gravitational attraction between objects increases as they are brought closer together (Newton's inverse square law). The escape velocity is the speed required to escape from a gravitational field. The Schwarzschild radius is the radius distance where the gravity is so great that the speed of light is required for escape velocity. Such velocity is impossible therefore the Schwarzschild Radius is where the force becomes strong enough to trap even light photons, and therefore is the radius of the black hole.

Continuity Laws.

A continuity equation is the mathematical expression in fluid mechanics that states; for the mass of fluid passing through a tube in steady flow, the mass flowing through any section of the tube in a unit of time is constant.

Conservation law.

The principal that a system quantity or property (e.g., matter, or space) remains constant during and after an interaction or process.

Conservation of angular momentum.

The total angular momentum of a system has constant magnitude and direction as long as the system is subject to no external force.

Conservation of linear momentum.

The principal that the linear momentum of a system has constant magnitude and direction if the system is subject to no external force.

Conservation of momentums.

The total momentum (angular and linear) of a system is a constant as long as the system is not subject to any external force.

Equivalence Principal.

The equivalence principal states that the force of gravity is indistinguishable from acceleration due to a changing rate or direction of motion. In 1917, Albert Einstein developed the Field Equations which include the equivalence principle. These combined with his general relativity theory express in mathematical terms the interaction of matter (mass and energy), radiation, inertial and gravitational forces.

Einstein's Laws of Relativity. *(Some still call this theory, yet these principles have been proven many times over, are repeatable and verifiable, therefore should now be known as **law**).*

Mass of an object increases relative to its rest mass with an increase of velocity. Einstein defines the relationship with the mathematical expression:

$$m = \{ m_0 / (1 - v^2 / c^2)^{1/2} \}$$

From this mathematical expression, other relativity mathematical relationships, like time dilatation can be derived. (Time dilatation has been proven by orbiting an atomic clock and comparing the orbiting time shift in orbit with earth time.)

Appendix 2.0
Numerical Values
Distance—Area—Mass—Energy

Astronomy Data are from the Bibliography References. New Universe Theory data are directly from calculations and deductions

Astronomy Data:

Earth to Sun	92,000,000 miles
Solar System Diameter	2 light years
Earth Orbit Distance	578,052,560 miles
Speed of Earth surface rotation at equator	1000 miles per hour
Speed of Earth around sun	65,954.3 miles per hour
Speed of Solar System around MWG	49,000 miles per hour
Solar System distance to center of MWG	28,000 light years
Time for solar system to orbit MWG center	~240 to 250 million years
Earth to moon distance	146,276 to 225,280 miles

Earth to Sun distance	92,000,000 miles
	8 light minutes
Solar System Diameter (including Oort cloud)*	18,200 Billion miles
	1.1 Light Years
Nearest Star (Centauri-Proxima)	4.2 light years
Nearest, most visible Open cluster	Pleiades
Pleiades star population	10,000 stars
Nearest Globular Cluster (M-4)	5,600 light years
Globular Cluster M-4 population	~100,000 stars
Globular cluster population range	75,000 to 1,000,000 stars
M-4 diameter (spherical)	5 light years
MWG habitable zone (est)	7 to 9 Kmps radius
Milky Way Galaxy (MWG) diameter	~100,000 light years (~30Kmps)
MWG thickness at solar system site	~15 light years
MWG thickness of disk (at sun's location)	3.5 Light Years
Galaxies seen by naked eye	Andromeda, MWG, Large & Small Magellan Clouds
MWG nearest spiral neighbor Andromeda	2.2 million light years
MWG Satellite Orbiting Globulars	150 to 200
MWG super-cluster (Local Group)	10 to 20 galaxies
MWG major group of Galaxies (Virgo)	250,000 (?)
MWG population	100 to 400 Billion Stars

Naked eye Quasar (#3C273)	From top of Pike's Peak
Andromeda and MWG closure speed	10,000 miles per hour
Andromeda distance	700 kiloparsecs 2.2 million Light Years
Andromeda time till collision w/ MWG	10,000 years
Andromeda Star population	400 Billion Stars
Abell Super cluster location	.8 to 2.5 BLY
Abell Super cluster direction	
Number of Galaxies in Abell Super Cluster	20,000

New Universe Theory data:

MWG to center of Universe distance	3.8 Billion Light Years
Universe Current Visible Diameter	34.4 Billion Light Years
Total Universe Diameter	78.8 Billion Light Years
Age of the Universe, today	78.8 Billion Years
Direction to center	opposite Abells

Equivalences:

Speed of light (fastest possible speed)	186,000 miles /sec
	299,793 (~300,000) Km /sec
One light year distance	5.87 Trillion miles
One Parsec =	3.261633 light years
One Mega-Parsec	.00326 BLY
1000 Mega-Parsec =	3.26 BLY (Billion Light Years)
One BLY	5.87 Billion Trillion miles
One Light Year =	9.44 Trillion (X 10^{13}) Km

Nuclear Physics Reference Data:

One MeV (Million electron Volts) =	4.45×10^{-20} Kwh
One Kwh (Kilo-watt-hours) =	2.24×10^{-19} MeV
One pound =	453.5924 gm (Grams
One Barn =	1.00×10^{-24} sq cm
One **A** (Angstrom) =	10^{-8} cm
Elementary charge =	1.602192×10^{19} coulombs
Electron rest energy =	.5109989 MeV
Positron rest energy =	.5109989 MeV
Electron rest mass =	9.109381×10^{-28} gm
Positron rest mass =	9.109381×10^{-28} gm
Proton rest mass =	1.672621×10^{-24} gm
Proton rest energy =	938.2719 MeV
Neutron rest energy =	939.5653 MeV
Neutron rest mass =	1.674927×10^{-24} gm
Neutron half life (at rest) =	10.25 minutes
Hydrogen Atom mass =	1.67335×10^{-24} gm

Appendix 3.0
Equations & Descriptions

Newton's Laws of Motion:
(Not listed in any particular sequence)

Force required for acceleration of any mass.

$F = m\,a$ *Newton's Law of Motion.*

Force equals mass (m) times acceleration (a).

Momentum, is mass times velocity. (= m v).
Velocity, is change of distance per unit time. (= ds / dt).
Acceleration, is rate of change of velocity. (= dv / dt).
Since mass and acceleration are both a function of velocity, The equation for Force is more accurately stated as rate of change of momentum. [d(m v) / dt].

$F = [\,d(\,m\,v\,)\,/\,dt\,]$.

A continuing force is required for a continuing change of momentum of any object (due to change of velocity, mass, or both). Near the speed of light we must consider Einstein's relativity, see equations later. (*Near the speed of light,*

Einstein shows time also to be a variable. That subject is beyond the scope of this book.) More accurately Newton's laws are modified to consider mass change with velocity, and therefore the law is:

$$F = [\, d(mv)/dt\,], = [\,(v\,\{dm/dt\}) + (m\,[dv/dt])\,].$$

When the mass of an object changes for any reason, (e.g., velocity enhancement at speeds near light speed; or, for a rocket, during take-off fuel burning reduces the rocket's mass). This equation is more definitive than $F = ma$, which is OK when velocities are far below the speed of light where the mass is almost the same as rest mass.

****

Positronium Stability Equations:

<u>Centrifugal Force (Fc)</u> is equal to the centripetal force, which is that force (string tension when a ball on a string is whirled) required to keep a mass (m) object (e.g., a ball) at a continuous velocity (v) and continuous speed (N) at a fixed distance (r) while revolving about a center of rotation.

$$Fc = m(v^2)/r \qquad \text{Newton's Law of Motion}$$
$$= [m(\pi^2)(N^2)r]/900$$

N = the rotation speed in revolutions per minute.
π = the constant 3.14159.

Force of Gravity (Fg) between two objects is equal to the product of the masses (m1 and m2) divided by the square of the distance between them.
$$Fg = k_1\,[(m1 \times m2)/d_2] \qquad \text{Newton's Law of Gravity.}$$

<u>Force of Electrostatic Attraction (Fq)</u> between two objects of opposite charge (q) is equal to the product of the charges

divided by the square of the distance (d) between them. Similarly, the electrostatic mutual repulsion force between two objects of like charges is the product of charges divided by the square of the distance between them.

$$F_q = k_2 [(q_1 \times q_2) / d^2]$$ Coulomb's Law

Positronium Stability

$$F_c = F_g + F_q$$

Combine (solve simultaneously) above four equations for Fq, Fc, and Fg, to determine values for N (rpm) as a function of d (centimeters) using electron/positron values for q and m.

****

Doppler Shift:

Doppler Red-Shift is defined as the change in the wave length ($\Delta \lambda$) of electromagnetic radiation (light), as a result of relative velocity (v) between the source and the observer. When the source is moving away the wave is lengthened, shifting it towards the red end of the spectrum. The frequency (ν) of that wave is also decreased. When the source is moving towards the observer, the wave length is shortened and the frequency is increased.

    Δ = change
    λ = wave length
    ν = frequency
    v = velocity of radiating source relative to the observer
    c = velocity of light (300,000 Kilometers per second)
    A = unit of length (10^{-8} Centimeters)

<u>Red Shift</u> effect is the increase or decrease in wave length due to the respective recession (or approach) velocity of

the wave source, when compared to the wave length from a source at zero relative velocity.

$$(Z) = [(\lambda - \lambda_0)/\lambda_0]$$
$$= [(\lambda/\lambda_0) - 1]$$

An example;
$$Z = [(7000A/3500A) - 1] = 1.0.$$

Examine the equations; as v approaches c, v/c approaches the value of 1, and the denominator approaches zero. Therefore, the value of Z approaches infinity (∞). For v/c = .5; $\lambda/\lambda 0$ = 2.0; and Z = 1.0.

In the real situation, "New Universe Theory" red-shift, the fact that the emitting object is not receding directly away from the observer, and the angle of recession must enter the equation. The red-shift tells us how fast the light source and receiver are separating, not their individual velocities. (In addition, relativity and time dilatation must also be considered for use of red-shift as a distance calculator. This is beyond the scope of this document). Time dilatation is not included in the nomograph of "z" values, velocity, % of light speed, and speed, verses distance with

The correct equation including and considering Einstein's special relativity for the value of z:
$$\Delta\lambda/\lambda = z = [(\{1+v/c\}/\{1-v/c\})^{\frac{1}{2}} - 1]$$

Which results in the equation for correct value of the component of velocity (v), that is directly away from the observation:

$$v = c[\{(z+1)^2 - 1\}/\{(z+1)^2 + 1\}]$$

<u>Frequency and wave length</u> are related by the equation; Frequency (v) = velocity (v) of the wave divided by the wave length (λ).

$v = v / \lambda$

Velocity for all electromagnetic radiation, it is the same as the speed of light (c), which is a constant at ~ approximately 300,000 Kilometers per second. (Actually the precise number is 299,728.377)

Further; An Einstein's Law of Relativity consideration is that red-shift is influenced. (Discussed in this appendix is the velocity enhancement of gravity). "Light can be red shifted by a gravitational field. (Verified by experiment according to Reference 2). Photons struggling to escape an intense field lose some of their energy in the process, and as this happens, their wave lengths are shifted towards the red". Gravitational fields near black holes have such strong influence that no light can escape. The extreme velocity-enhanced-gravity fields during, and immediately following the STAGE II processes must therefore, also cause some red shift. This type of shifting is referred to as non-cosmological, which indicates that it is not simply due to velocity alone. It is not known how to correct distance (and time) for this portion of measured red-shift.

****

Hubble Lines:

<u>Hubble Number (H)</u> "constant" was initially defined by Edwin Hubble and Milton Humason as velocity divided by distance (d). We now know that this relationship is only approximately valid, and then deviates more from a straight line as velocities approach and exceed 30% of c (velocity of the speed of light). It is defined as velocity (Kilometers per second) per unit of distance (in Mega-parsecs).

$$H = v / d$$

Hubble "number" was originally called a "constant". Astronomers have measured "Hubble numbers" in various directions and the numbers range from 50 to values well over 100. The number changes with distance as well with distance. (See figure 10.3)

A Hubble number measurement/calculation for red–shift according to Liz Gebis, Seeran Mittappali, and Michael Choi; They jointly and recently measured the red–shift and calculated the Hubble number in the direction and range of the Abell Galactic clusters at 117+. Their red-shift correction is: $[(r^2 + 2r) / (2 + 2r + r^2)]$. (Internet published by NASA in APOD)

****

Volumes:

Volume of Local Universe:
    @ 1 Bly radius = 33.811 cubic BLY

Surface area of Universe:
    @ 39.4 Bly Diameter = 4876.9 square Bly
    @ 78.8 Bly Diameter = 19597.5 square Bly

Volume of Deflagration Front (Stage II):
    @ 39.9 Bly Diameter, = 12,192.3 cubic Bly
    @ 78.8 Bly Diameter, = 48,993.8 cubic Bly

Reference info:
    Spherical Volume = $4/3\pi r^3$.
    Sphere Surface Area = $4\pi r^2$.
    Sphere Shell Volume = $4/3\pi [(r_1)3 - (r_2)3]$.
    where: π is 3.14159,
    (r_1) is major radius,
    (r_2) is minor radius.

****

Entropy: and, Velocity Reductions in Fluid Flow:

Entropy Increase = (velocity and energy reductions)

Entropy is unrecoverable energy.
Fluid flow relationships:

Laminar Flow; $v = (\Delta p)/(4 \mu L) \times (R_0^2 - R2)$
$\Delta P = kV$

Turbulent Flow; $v^2 = ((\Delta p) D \times 2gR)/L$
$\Delta P = kV^2$

Reynold's Number; $R_n = \rho V L / \mu$
$R_n < 2000$ yields laminar flow
$R_n > 4000$ yields turbulent flow
$R_n > 2000$ and < 4000 is transition

Entropy Increase, $E_1 \propto \Delta p$; Function of Velocity decrease

Viscosity, μ = The resistance to flow between layers within any fluid.

****

Relativity; Velocity Enhancement.
Mass & Force of Gravity:

Einstein's 'theory' of relativity!. Until recently, it was called theory, but since it has been proven as fact, henceforth it is upgraded to: "Einstein's <u>Law</u> of Relativity".

As written by Einstein: $E = mc2 / [1 - (v2/c2)]^{½}$

Restated: $m = m_0 / [1 - v^2/c^2]^{1/2}$
Also: $m/m_0 = 1 / [1 - (v^2/c^2)]^{1/2}$

I have found that the most illustrative way to show the velocity relativity to enhanced mass and gravitational force is with a table of v / c along-side the two ratios. Simply by assigning values between 0.0 and 1.0 and proceeding through the calculations to determine the values of m divided by m0. Gravitational force enhancement is simply a square function of the mass enhancement. (Two adjacent objects with the same rest mass and traveling at the same velocity attract each other by Newton's Laws of Gravity).

$$Fg = [(m1 \times m2) / d^2]$$

A table illustrating the velocity enhancements of mass and gravity is presented in the text. Examine the equations, and the enhancement significance is seen by observing that both m and Fg approach infinity (∞) when v approaches c, and they both approach 1.0 when v approaches zero (0). A summary table is shown:

v/c	m/m_0	Fg/Fg_0
0.00	1.00	1.00
0.25	1.03	1.07
.50	1.15	1.33
.75	1.51	2.28
.99	7.09	50.
.9999	70.70	5,000.
.999999	707.0	500,000

****

Appendix 4.0
Bibliography
(References and Additional Reading)

(1) **Our Universe**, Presentation by Professor of Astronomy at the University of Washington, Bruce Margon, about 1998. The presentation was televised on The Education Channel in Phoenix AZ, on Cox Cable, .January 2002

(2) **"Exploring the Dynamic Universe"** Astronomy Text by Theodore P Snow; Professor of Astronomy at the University of Colorado at Bolder. Published by West Publishing Company, St Paul MN. Copyright 1988, rev 1995.

(3) **Element formation in the Big-Bang**, based on diagram Of Ref (1) and on Data from R V Wagoner, The Astrophysical Journal, 1973.

(4) **Mapping the Universe,** Presentation; by Margaret Geller and John Huchra on their survey project, presented to the Astronomy Club of Sun City West on November 2001.

(5) **The First Three Minutes"** Steven Weinberg, Professor of Physics at Harvard University ; (Professor of Physics

at University of Texas in 1993), published by Basic Books of the Perseus Books Group. Copyright 1977: second edition 1993

(6) **Hubble's Constant and the expanding Universe,** W. Freedman, Carnegie Observatories, HST Key Project Team, and NASA,. Distance vs Velocity Graph of Abell Clusters. Y axis is velocity, and X axis is distance. ($Y = 117.67X - 4728.3$) Velocity in kilometers per second vs Distance in MegaParsecs, 2001.

(7) **"General Chemistry, Principles and Structure"** by James E. Brady, St. John's University, Jamaica, New York; and; Gerard E. Humiston, Skidmore College, Saratoga Springs, New York. Published by John Wiley and Sons, New York, 1986.

(8) **"Galaxies"** (coffee table book) by Timothy Ferris, Professor emeritus at the University of California at Berkeley. Published by Harrison House, distributed by Crown Publishers, Inc. 1987

(9) **"Cosmos"** by Carl Sagan, was Professor of Astronomy and Space Sciences at Cornell University. Book based on 13 part television series, Copyright 1980, Carl Sagan Productions, Inc.

(10) **"An Introduction to Astronomy"** by Robert H. Baker, Ph D., University of Illinois, Published by D. Van Nostrand Company, Inc., New York, 1947.

(11) **"Physical Geology"** by Charles C Plummer and David McGeary, William C Brown; Publishers, 1988.

(12) **"The Handy Space Answer Book"** by Phillis Englebert and Diane L Dupuis, Visible Ink Press, 1998.

(13) **"Galileo's Daughter"** by Dava Sobel, Sorbel acquired much of the correspondence between Galileo and Galileo's daughter (a Nun), which reveals much of Galileo's successes and woes. Penguin Books, 2000.

(14) **"Astronomy and Cosmology, A modern course"** by Fred Hoyle, Professor at California Institute of Technology, Pasadena California. Published by W H Freeman Co, 1975.

(15) **"Mechanics___Statics and Dynamics"** by Merit Scott, Professor State College, Pa. Published by McGraw-Hill Book Co, 1949.

(16) **"Astronomers"** Companion book to PBS Television Series, by Donald Goldsmith, Published by St Martin's Press, 1991.

(17) **"Stars and Planets"** by Jay M Pasachoff, Houghton Mifflin Co, 2000.

(18) **"Stephen Hawking's Universe"** by David Filkin, Basic Books, 1997.

(19) **"The Creation of the Universe"** by George Gamow, Viking Press, 1952.

(20) **"A I P Style Manual"** by American Institute of Physics Publication Board, A I P, 1997.

(21) **"Supernovae ...."** Article by Lifan Wang & J Craig Wheeler, Sky & Telescope Magazine, January 2002.

(22) **Orion Nebula,** Image of, by Mark McCaughrean, Joao Alves, Hans Zinnecker, and Francesco Palla, Published in Sky & Telescope magazine, February 2002.

(23) **"Local Universe"** Illustration by Elizabeth Rowan in article by Free Lance Writer Steve Nadis, published in Astronomy magazine, March 2002.

(24) **"..White Dwarfs...pin down cosmic age..Date the Universe.."** Articles by Vanessa Thomas, and by Alan MacRobert, Free Lance writers. Respectively; Astronomy magazine, August 2002, and in Sky & Telescope magazine, August 2002. Same subject was briefly discussed on APOD.

(25a) **"Do You Believe in the Big Bang?"** Article by Jim Sweitzer, Director of Space Science Center at DePaul University in Chicago, Published in Astronomy Magazine, December 2002.

(25b) **"Five reasons why you should believe the Big Bang"** Quote by Rick Fienberg, editor in chief, Sky & Telescope Magazine, published July 2002.

(26) **"Electricity and Magnetism"** by Norman E Gilbert, Visiting Professor of Physics, Rollins College, Also Dartmouth College, Published by The Macmillan Company, 1950.

(27) **"Webster's Encyclopedic Unabridged Dictionary of the English Language"**, Gramercy, Random House Publishing Copyright C 1997.

(28) **"Introduction to Nuclear Engineering"**, by John Lamarsh, Addison-West Publishing Company, C 1977.

(29) **"Encyclopedia of Science and Technology"**, Joe Faulk editing manager, Roger Kasunic director of editing, design, production. Published by McGraw Hill, 20

volumes; Eighth Edition, copyright 1960-1997, New York McGraw-Hill Book Co.

(30) **"Distance verses Velocity, for Abel clusters"**, Choi, Gebos, Metsopoli. Astronomical researchers. Published by APOD, 2000 . (Authors home pages: emgebos@amsi.edu; ramrom@imsa.edu; and mchoi@imsa.edu.)

(31) **"the extravagant universe"** by Robert P Kirshner, Clowes Professor of Science at Harvard University and Head of the Optical and Infrared Division at Harvard-Smithsonian Center for Astrophysics. Published by Princeton University Press, 2002.

(32) **"Faster Than The Speed Of Light"** by Joao Magueijo, Professor of Physics at Imperial College, London England. Published by Perseus Publishing, 2003.

(33) **"the black hole at the center of our galaxy"**, by Fulvio Melia, Professor of Physics and Astronomy, Head of Physics and Astronomy Department, Arizona University. Published by Princeton University Press, 2003.

(34) **"Before the Big Bang"**, by Ernest J. Sternglass, Professor of Radiological Physics at University of Pittsburgh. Published by the Four Walls Press, 1997.

(35) **"Red Giants and White Dwarfs"**, by Robert Jastrow, Director, Goddard Institute for Space Studies. Published by Harper and Row, 1967.

(36) **"The Red Limit"**, by Timothy Ferris, Professor Emeritus at the University of California at Berkeley, published by Perennial Press, 1983

(37) **"A Visit to Mars Hill"**, by David Healy, article contributed to Astronomy Magazine, June 2004.

(38) **"The Enigma of Przybylski's Star"**,. by Guy Worthey, assistant professor of Astronomy at Washington State University; and Charles R. Cowley, professor of astronomy at the University of Michigan. Published by Sky & Telescope, August 2004.

(39) **"Globular Cluster Systems"**, by Keith M. Ashman, Department of Physics and Astronomy, University of Kansas; and Stephen E. Zepf, Director of Astronomy, University of California, Berkeley. Published by Cambridge University Press, 1998.

(40) **"The Physics of Neutron Stars"**, by J. M. Lattimer and M. Prakash, Department of Physics and Astronomy, State University of New York, Sony Brook, N. Y. Published by AAAS journal Science, 23 April 2004.

(41) **"A Possible Quark-Star Discovery"**, by Jack Lucentini, Sky&Telescope Magazine, July 2002.

(42) **"The Hunt for Dark Matter in Galaxies"**, Ken C Freeman, Mount Stromlo Observatory, Australia National University, Weston Creek, ACT 2611 Australia. Published in AAAS journal 'Science', 12 December 2003.

(43) **"Cosmology's Treasure Map"**, Govert Schilling, Contributing editor for Sky and Telescope magazine, Published February 2003.

(44) **"Moving Right Along"**, Mario Livio, Head of Science Division of Space Telescope Science Institute, in Baltimore. Printed in Astronomy Magazine, July 2002.

(45) **"The Biggest Thing ever Found"**, Govert Schilling, published in Sky and Telescope Magazine, February 2004.

(46) **"American Heritage dic-tion-ar-y"** fourth edition published by HougtonMufflin, 2004.

(47) **"Wikipedia, The free encyclopedia"**. An on-line encyclopedia started in 2001. (Http://en.wikipedia.org/wiki/main.page)

(48) **"Explorers of Mars Hill"**; 1894 + 1994, William Lowell Putnam and Others, 1994. Phoenix Press, Maine.

(49) **"Gravity from the Ground Up, An Introductory Guide to Gravity and General Relativity"**. Bernard Schultz, 2003, Cambridge University Press, Cambridge, Mass.

(50) **"Introduction to Fluid Mechanics"**, Stephen Whitaker, 1968, Krieger Publishing Co., Malabar, Florida.

Appendix 5.0
Acknowledgments:
(no particular sequence)

Thanks:

Nancy McGehee, *Spouse, Humanitarian.* Writing consultant, and continuing loyal support including picking up the slack in our lives, especially during the long hours, days, and calendar time, which allowed me the time to develop and write, and prepare this manuscript for revealing the New Universe Theory.

Bobby McGehee Jr, *Son, Technician Boeing Flight Test.* For the time and thoughtful contributions to early draft proof reading including the grammatical and human sensitivity editing. An essential contribution. His life long history of extensive science fiction book reading makes him an important contributor

Patricia Suzanne (Sue) Hart, *Daughter Psychology major.* For the unsolicited, thoughtful written message which I used for the "Introduction". When she wrote her letter she had no idea her letter would be used for this. Also for critiquing the manuscript for presentation sensitivity.

Ron Mobley *(Pres Sun City West Astronomy Club)* **and Sue Mobley**. Special friends who remain loyal and tolerated my evasiveness when inquiring about my book's subject matter during NUT analyses and manuscript preparation. For reading and critiquing the manuscript. For book cover and other presentation ideas.

William "Bill" Ryder, *Nephew, Career Nuclear Engineer.* Palo Verde Nuclear Power Plant, Graduate Oklahoma University. For consultation support with nuclear engineering calculations, and for use of his nuclear engineering reference books.

Monnie Ryder, *Niece, Petroleum Engineer, Masters in Business Management.* For contributing with her Technical Writing expertise.

Donald Holmlund, *Friend, amateur Photographer.* Photographs of Washington "Figs", 2002.

Paul Turley, *Graphic artist,* for perseverance with computer graphics support by converting my engineering pad drawings and graphs into publishable graphic figures.

Lindsey, Ashley, and Monnie Ryder for computer graphic illustrations for the book covers. Lindsey and Ashley are 12 year old computer whizzes.

Grant Thompson, *Art graphics,* Sun City West, AZ.

Appendix 6.0
Author's Credentials'

The Author, Bobby L McGehee is a Career Engineering Physicist. He studied, reviewed, and reinterpreted the extra-galactic red shift, to explain universe expansion phenomena consistent with the Laws of Physics. His thinking with scientific logic concluded with development of the **New Universe Theory**. His **new theory** opens the door for redefining astronomy and cosmology as true sciences; **without magic or myth.** This theory includes descriptions of **"Primordial Matter and the Reduction Mechanisms"; and is totally consistent with the Laws of Physics.** This NUT first was thought of as the **Development** of a New Concept, but now, it seems more appropriately described as the **Discovery** of the Origin of the Universe, ...**but what is the difference?.**

Bobby L McGehee, was born in Enid, Oklahoma on September 19, 1927, and graduated from Enid High School in 1945. Then after two years in the USMS and the USAF, he entered Oklahoma State University and graduated in 1952. He qualified for three degrees; Engineering Physics, Physics, and, Education. In his fifth year while taking graduate courses in physics, he taught basic physics as an assistant to the Physics Professors. While in college he accumulated over 220 semester credit hours (equivalent to 293 Quarter Hours) in various fields of study. After 32 years as a Career Engineering Physicist he returned to college for another three years, studying Geology and Astronomy.

Bobby L McGehee began his Engineering-Physicist Career in 1952 when he was employed at the Beach Aircraft Corporation, but was at The Boeing Company for most of his career. During this time he developed and designed aircraft research and development test facilities, including Jet Engine Test Stands, Low Speed, Transonic, and Supersonic Wind Tunnels, Environmental and Anechoic Acoustic Test Chambers. On two occasions out of an engineering staff of several thousand, he received awards for Engineer of the month and later for Engineer of the Quarter. For organizational contributions he was awarded a Certificate of Recognition by the American Institute of Aeronautics and Astronautics, where he was an Associate Fellow. He also served as an officer on the board. He wrote numerous engineering reports and published technical papers on "Transonic Wind Tunnel Designs for testing at transonic speeds to 1.2 Mach" and another on "Transonic Wind Tunnel Design for testing to Mach 1.3". He wrote, published, and presented a technical paper at the national meeting of the Institute of Environmental Sciences on the design of the Boeing Large Anechoic Test Chamber, "A Test Facility for Aircraft Jet Noise Reduction". This paper was rewritten and published in three languages in the international Journal for the **ICAO** (International Civil Aviation Organization). He earned patents on a Jet Engine Noise Suppressor, and shared the patent for inventing a counter rotating turbine drive that makes possible small scale model counter-rotating Un-ducted Turbofan testing.

He continues active in Astronomy organizations and meetings. He is a member of the American Association for the Advancement of Science (**AAAS**), American Association of Physics Teachers (**AAPT**), and is an avid reader of Science Journals, Nature, and other publications on subjects relating to astronomy and cosmology. He is a member of the Astronomy Club of Sun City West, Friends of Lowell Observatory, the **ASP** (Astronomy Society of the Pacific) and when time allows, attends national astronomy meetings.

And now Bobby McGehee has taken on the biggest project of all, The Universe and Beyond.

Appendix 7.0
Tributes:

Sir Hermann Bondi
Sir Thomas Gold
Sir Fred Hoyle
and
Grote Reber.

For the example they set, and for their honorable and relentless scientifically credible adherence to the Laws of Physics; To show them the respect they deserve, but never received during their lifetimes; brief tributes are presented to these scientifically credible Giants. I believe they all very much deserve and if they were all still alive, would be pleased and would appreciate being associated with this New Universe Theory with the Laws of Physics.

<u>Hermann Bondi   1919—20   .</u>
<u>Thomas Gold   1920—2004.</u>
<u>Fred Hoyle   1915 —2001.</u>
Men of high ethical and scientific standards. Although frequently under pressure from colleagues, These true

scientists never yielded commitment to the standards of scientific thinking processes. They scrupulously examined all of the concepts then available for the origin of the universe, and the BB idea was obviously (understood by them) inconsistent with scientific principles and violated Laws of Physics. These men developed a scientifically credible theory for the origin of the universe, but it simply did not happen that way, (and they mis-used the word 'creation' in place of 'generated'). Some other astronomers ridiculed, and some scoffed at their ideas (ideals) for not agreeing with the generally accepted BB theory. Some said it was proven; But these men knew better, and now, we all know better. They yielded their support for the steady-state theory after the discovery of quasars. But disproof of steady state was not reason for acceptance of another, even faultier theory, namely the BB, which will be remembered in infamy.

Many writers and astronomers honored the deceased with obituaries, but subtly apparent, many were condescending as they still by their skewed comments indicate disapproval for disbelief in the BB. Some imply that they were stubborn. Maybe so, but what really was, is steadfast in belief and loyalty to the Laws of Physics. They stuck with integrity and with scientific principles to the end.

And now Hoyle and Gold rest in peace, because their principles were right, and they knew they were right. We all now know they were right. This New Universe Theory is not only a revelation, but also vindicates, and re-establishes their stature among the greatest few. These were great men, and undoubtedly, these men were the more scientifically credible professional astronomers among all of the many great astronomers of the 20th century. From character and respect for their peers, we can conclude that they would forgive all of the dissenters from scientific principles, in particular, the BB supporters who knew not what they didn't know.

There have been many well deserved respectful tributes written. Many were found on the internet and in many 'letters to the editors' of Astronomy Magazine, Sky & Telescope Magazine, and in other scientifically oriented publications. It is worth your time, just fill in the search window on your home page with the name "Hermann Bondi", "Thomas Gold", or "Fred Hoyle".

Grote Reber   1911 -- 2002.

This man is the pioneer in radio astronomy. He recognized early on, that the electromagnetic radiation would be 'red-shifted' completely out of the optical range. He erected the worlds first radio telescope (31 foot dish) in 1937 in his back yard in Wheaton, Illinois. For almost a decade he was the world's only radio-astronomer. Radio waves also made it possible to map out the spiral arms of the Milky Way Galaxy. He opened what is now this major field of astronomy.

Grote Reber moved to a remote area in Tasmania, Australia where there was less electromagnetic interference from man made radio signals, as well as a good view of the southern sky. He constructed radio antennas in a grouping that was about 3,500 feet in diameter on his 223 acre private property.

He was a graduate electrical engineer from Illinois Institute of Technology, and was not trained as a physicist or as an astronomer. However, in 1962 he received an honorary Doctorate from the Ohio State University.

Dr. Grote Reber was an explorer and he never believed in the Big-Bang theory. Reber believed in real, provable phenomena, and had no use for imagined ideas that were inconsistent with the Laws of Physics. Professor of Astronomy at Wheaton College Joe Spradley, said that Reber gave a lecture titled "Big Bang Is Bunk". Grote Reber tried to develop an alternative, but was never successful. However, he was successful in maintaining his scientific credibility and integrity, throughout his life. (Many of the above statements for Grote Reber are excerpts from an obituary by Rex W. Huppke, Chicago Tribune, Dec 20, 2002; they are reprinted with permission).

This New Universe Theory didn't get published in time, but nevertheless, Sir Thomas Gold, Sir Fred Hoyle, and Dr Grote Reber alike, are now more than vindicated, they were humble, and will now deservingly honored in history.

Appendix 8.0
Concluding Comments

It has been refreshing to note that since about 1975, most physicists are concluding that the BB standard model falls short of answering an increasing number of important questions: Why does the universe contain so much stuff? From where did all this stuff come? Where is and what is the invisible mass that dominates the universe? All objects in the universe are revolving and rotating, so what is the energy source for all of this angular momentum? Why are Galactic clusters and super-clusters accelerating away from each other? (My conclusion; they aren't).

Also, it has been perplexing and bewildering to find almost every thing I read, cosmology or astronomy related, either directly or indirectly, almost always, refers to the Big Bang theory, as if it were unquestionably real. I often add a note in the borders of most articles: "More BBB". This includes articles in quality publications; e.g., "Astronomy", "Discover", "National Geographic", "Nature", "Science", "Scientific American", "Sky & Telescope", etc., and text books as well.

\My hope is that present and future generations will continue to ponder, question and analyze with scientific integrity by using all available information, any theories and discoveries presented. Inquiring minds and curiosity will benefit mankind.

ABOUT THE AUTHOR

Career Engineering-Physicist Bobby McGehee was born in Enid Oklahoma, studied at Oklahoma State University, qualified for degrees in Engineering-Physics, Physics, and Education. During his exciting 32 year career in the aircraft industry he was honored with Engineer-of-the-Month and Engineer-of-the-Quarter awards, was an Associate Fellow of the American Institution of Aeronautics and Astronautics, and also served on the Board. He later returned to college to study Astronomy and Geology. He is an avid reader of scientific discoveries and regularly attends local and national astronomy society lectures. Since 1950 he has queried the BB theory in an attempt to prove it to himself. His persistent questioning led him to develop the Laws of Physics compliant New Universe Theory. (Credentials in Appendix 6.0.)

ABOUT THE BOOK

Big-Bang? Do you believe it? I don't. Being a career engineering-physicist, I always look for proof. For 50 years I have tried to prove the BB to myself. I have reviewed Slipher's red shift observation, reinterpreted it, and corrected Hubble's declaration, (the universe is not expanding and exploding to its death, instead, it is growing with vim and vigor). Red shift defines velocity, not acceleration or deceleration. So, I developed the New Universe Theory which is believable because it complies with known facts and the Laws of Physics. I am convinced this is the true origin of the universe. I do not know the origin of primordial matter.

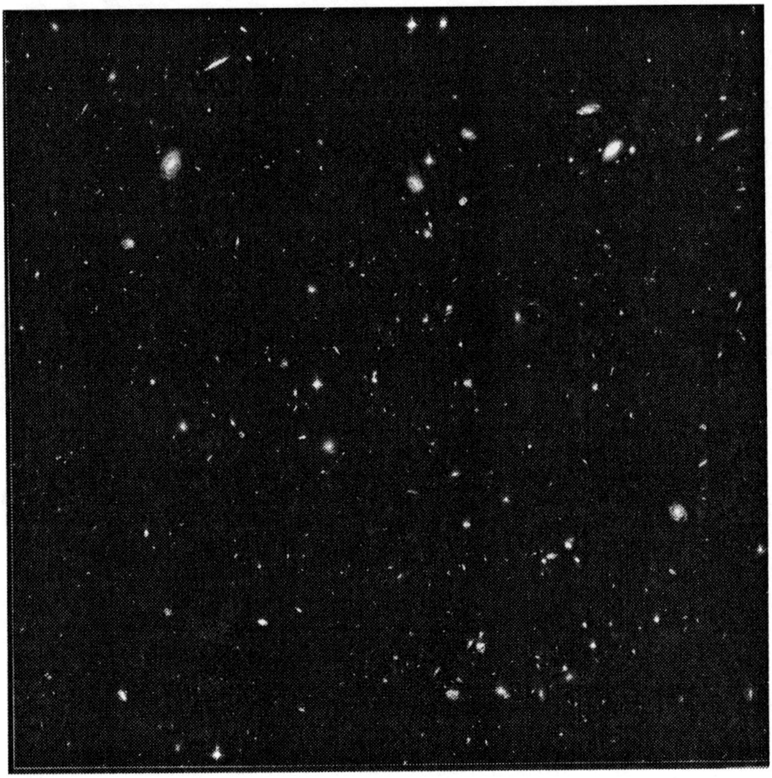